CONSTRUCTION PROJECT SAFETY

CONSTRUCTION PROJECT SAFETY

John Schaufelberger
Ken-Yu Lin

WILEY

Cover design: Michael Rutkowski

Cover photograph: © Nic Lehoux/The Northwest Studio Inc.

This book is printed on acid-free paper. ♾

Published by John Wiley & Sons, Inc., Hoboken, New Jersey.

Published simultaneously in Canada.

Library of Congress Cataloging-in-Publication Data:Schaufelberger, John, 1942-

Construction project safety / John Schaufelberger, Ken-Yu Lin. – First edition.

pages cm

Includes index.

ISBN 978-1-118-23192-0 (hardback); ISBN 978-1-118-41951-9 (ebk); ISBN 978-1-118-42100-0 (ebk)

1. Building–Safety measures. 2. Construction industry–United States–Safety measures. 3. Construction industry–Health aspects–United States. 4. Construction industry–Safety regulations–United States. 5. United States. Occupational Safety and Health Administration. I. Lin, Ken-Yu, 1975- II. Title.

TH443.S335 2013

690'.22–dc23

2013023530

Printed in the United States of America

SKY10031758_120221

CONTENTS

PREFACE

Construction continues to be one of the most dangerous occupations in the United States, accounting for about 10% of the disabilities and 20% of the fatalities that occur in the industrialized work force. Most successful construction companies have recognized that safety and health management is a critical strategic issue and have developed comprehensive company safety and health programs. Anyone involved in managing a construction project must understand that job site safety is very important requiring their full attention.

A safe working environment results in increased worker productivity and reduces the risk of injury. Accidents are costly, leading to disruption of thc construction schedule and demanding significant management time for investigation and reporting. The indirect costs that result from accidents can be significant and generally are not covered by insurance, thus adversely affecting the profitability of a project.

Project labor cost is greatly influenced by a construction company's safety record. The firm's workers' compensation insurance premium rates are directly related to the company's safety record. Federal and state statutes mandate the implementation of required safe working practices on all construction sites. Failure to comply with these requirements may result in significant fines as well as accidents.

This book was developed to examine construction project safety from the typical phases of a construction project. It was written for use in undergraduate and graduate construction management programs. Most authors approach construction safety from a regulation-based perspective. It was felt that providing a project context for the discussion would enhance student learning regarding how to develop a site-specific accident prevention program by thinking through construction processes and identifying the various hazards that may exist during the different phases of construction.

Fire Station 39 constructed in Lake City, Washington, was selected as the project to be used as a case study throughout the text and serves as the basis for the discussion. The project was a $3.5 million fire station constructed between April 2009 and March 2010 for the Seattle Fire Department. The project was not so complex that students would need to learn complex construction techniques but instead can focus on project site safety issues.

The book is organized into individual chapters that address major construction safety and health issues. Chapters 1 through 4 discuss the importance of good job site safety procedures, workers' compensation insurance, accident prevention programs, and the federal Occupational Safety and Health Administration (OSHA) requirements. Chapters 5 through 10 discuss the safety concerns that were addressed during the various phases of constructing Fire Station 39. Each chapter concludes with a set of review questions that emphasize the major points covered in the chapter. Exercises are provided that require application of the principles discussed. Appendix A contains a glossary of terms used in the book, Appendix B is a sample accident prevention program for Fire Station 39, Appendix C contains a construction schedule for Fire Station 39, Appendix D contains the requirements for the 10- and 30-hour OSHA certification, and Appendix E contains selected OSHA standards that relate to Fire Station 39. Selected OSHA standards are contained in Appendix E to provide a reference for students to understand the safety plans discussed in the book. Even though some states may issue their own occupational safety and health regulations, these regulations generally are based on OSHA requirements.

An array of additional resources for both students and teachers are readily available on the book's dedicated web site, www.wiley.com/go/constructionprojectsafety. While the construction photographs and building information model images presented in the book are black and white, color versions of both the photographs and images can be found on the web site as well as selected construction drawings for Fire Station 39. An instructor's manual containing answers to the review questions and PowerPoint slides is available for teachers on the book web site.

Finally, the authors would like to thank The Miller Hull Partnership and the following individuals for their assistance in the completion of the book: Alan Sutherland, Jason Solie, Caroline Kreiser, Kyle Hughes, and Aran Osborne.

CONSTRUCTION PROJECT SAFETY

1

INTRODUCTION

1.1 NATURE OF CONSTRUCTION OPERATIONS

Other than mining and agriculture, construction project sites are the most dangerous workplaces in the United States. The construction industry accounts for about 10% of the disability injuries and 20% of the fatalities that occur in the industrialized work force, even though construction workers account for only about 8% of the industrial man-hours worked in the United States. The construction industry includes a wide variety of companies, specialized crafts, and types of projects. Projects vary from single houses to multibillion-dollar major infrastructure projects.

Fatality and recordable injury and illness cases within the industry from 2007 to 2010 are shown in Table 1.1. Therefore, construction safety is a very important topic. The ability of a construction company to eliminate or mitigate the risk of accidents is essential in the execution of a successful construction project. Implementing effective safety measures reduces project costs and demonstrates concern for the welfare of

Table 1.1 Construction Workplace Fatalities and Injuries/Illnesses per 100 Full-Time Workers

	2007	2008	2009	2010
Fatalities				
Number of fatalities	1239	1016	879	802
Rate of Injury and Illness Cases per 100 Full-Time Workers				
Total recordable cases	5.4	4.7	4.3	4.0
Cases involving days away from work, job restrictions, or transfer	2.8	2.5	2.3	2.1
Cases involving days away from work	1.9	1.7	1.6	1.5
Cases involving day of job transfer or restriction	0.9	0.7	0.7	0.6

Source: Bureau of Labor Statistics, "Industries at a Glance," available: http://www.bls.gov/iag/tgs/iag23.htm.

people working on or passing by the project. Safety records are often considered by project owners when selecting construction firms to construct their projects.

There are two major aspects of project site safety: (1) *safety of persons working on the site* and (2) *safety of the general public who may be near the project site.* Both aspects must be addressed when developing project-specific safety plans. These plans identify all hazards to be faced by construction workers and the general public during the various phases of construction and measures to be taken to minimize the risk of injury to workers or the public. Examples of construction worker safety measures are requiring the wearing of personal protective equipment (PPE) and placement of barricades around floor openings. Examples of public safety measures are perimeter fences and warning signs to prevent entry into the project site by unauthorized individuals.

The primary causes of construction job site injuries are:

- Falling from an elevation
- Being struck by something
- Trenching and excavation cave in
- Being caught between two objects
- Electrical shock

Many hazards exist on all construction sites: sharp edges, falling objects, openings in floors, chemicals, noise, and a myriad of other potentially dangerous situations. Mitigation measures are required to minimize the potential for injury, and continued training is needed to ensure the entire work force maintains a work safely attitude.

Most construction projects are unique and executed in varied work environments. Construction workers, therefore, are constantly expected to familiarize themselves with new situations that potentially may be hazardous. In addition, the composition of construction project teams varies from project to project, and many craft workers may work for different employers leading to a lack of conformity and continuity. Craft workers may only work on a project site during certain phases of the work and then move to another project site. The continuing change in the composition of the work force on a project presents significant leadership challenges to the project manager, superintendent, and field supervisors.

Another major safety challenge for construction site supervisors is the increased employment of workers for whom English is a second language. Not only do these workers have difficulty reading and understanding safety signage, but they may be unwilling to report unsafe job site conditions or working practices. It is critical that the supervisors be able to enforce good safety practices among all individuals working on a job site. This may require that safety signage be posted in multiple languages and that safety orientations be conducted in multiple languages.

As construction projects are being executed, there is a continuous series of situations in which construction workers and/or the general public may be exposed to risk of injury. It is extremely important for construction leaders to recognize these situations and take action to control or mitigate these job site hazards. Many construction operations occur in excavations below the surface of the ground or in the air above the ground. In many cases, construction activity is exposed to natural

Figure 1.1 Masonry workers construct a concrete block wall from a vertical scaffold

elements, such as rain, snow, wind, or other climatic conditions. Implementing measures to protect workers and the public is the best way to minimize the potential for injury. This is illustrated in Figure 1.1, which shows construction workers constructing a concrete masonry wall from a vertical scaffold.

Creating a safe work site is a function of the physical conditions of the working environment and the behavior or working attitude of the individuals working on the site. Safety planning must begin during the initial planning for a construction project along with the development of a cost estimate and project schedule. The initial safety plan needs to outline how safety will be managed on the project, including roles and responsibilities of project participants, resources available, anticipated hazards and mitigation measures, training requirements, and safety equipment needs. Requiring everyone on the project site to wear appropriate PPE may impact worker productivity, and the purchase of appropriate safety equipment may impact project costs.

1.2 IMPORTANCE OF SAFE PRACTICES

A disabling injury or a fatal accident on a construction project will have a significant negative impact on the execution of construction operations. Accidents cost money, have an adverse impact on worker morale and productivity, and lead to adverse publicity about the project, the construction company, and the project owner. It is the construction company's responsibility to provide a safe working environment for all construction workers on the project site, including those employed by subcontractors, and to protect the public from harm. This is a significant concern when major construction activity occurs within a facility, such as a hospital, that is in operation.

The primary factors that motivate safe practices on construction sites are:

- Humanitarian concern for workers and the public
- Economic cost of accidents
- Regulatory requirements for work site safety

Each of these factors is discussed in the following paragraphs.

It is a normally accepted principle that an individual should not be injured while working for an employer. This is based on humanitarian concern for the well-being of every individual. In addition to the humanitarian concern, there is a significant adverse economic impact if an accident occurs. Accidents are costly, as will be discussed in the next section, and often result in uncompensable delays in the completion of the construction project.

Congress passed the Occupational Safety and Health Act (the OSH Act) in 1970 establishing mandatory workplace safety and health procedures. These required procedures will be discussed in greater depth in Chapter 4. The act created the Occupational Safety and Health Administration (OSHA) within the Department of Labor to administer the act. OSHA regional and area offices employ inspectors whose duties include visits to construction projects to ensure compliance with mandated safety and health procedures and to assess significant fines for failure to comply with the required procedures. Job safety and health requirements generally consist of rules for avoiding hazards that have been proven by research and experience to be harmful to personal safety and health.

The act authorized individual states to establish their own occupational health and safety requirements as long as the state requirements are at least as effective as the federal requirements. Several states have enacted their own occupational safety and health statutes and employ inspectors to ensure compliance on construction project sites within the state. Failure to comply with statute requirements usually results in significant citations and fines.

Most successful construction companies have recognized the importance of safety management and have developed effective company safety programs that include:

- New employee orientation
- Safety training
- Project-specific accident prevention plans
- Job site surveillance

Good safety practices reduce the cost of doing business because they lead to reduced premiums for workers' compensation and liability insurance and minimize the costs that result from accidents and injuries on a job site. Construction companies, depending upon the type of craft labor that they employ, often pay 10 to 20% of their direct labor costs for workers' compensation insurance premiums, which is a significant cost of doing business. This will be discussed in more detail in the next chapter.

The effectiveness of a construction company's safety program often is a key factor in the ability of the company to become prequalified and allowed to submit a proposal on a project. Project owners do not want unsafe contractors working on their projects, because the owners do not want the negative publicity associated with construction accidents. Unsafe project sites also often lead to citations and resulting fines from state or federal occupational safety and health inspections.

Implementing and enforcing a strong safety program also demonstrate company management's interest in the welfare of individuals working on the job site. The ability of a construction company to deliver a quality project is directly affected by the ability and motivation of the individuals working on the job site, whether they work for the general contractor or for one of the subcontractors. Providing a safe working environment demonstrates management's commitment to the welfare of the workers resulting in the workers wanting to work on the project site and making the project a success.

The DuPont Company has been a leader in developing a safety culture within the company and provides consulting services to companies and organizations that wish to improve their safety performance. Based on their extensive experience with workplace safety, DuPont has developed the following 12-element program[1] for management of employee safety:

- *Management Commitment.* This is a basic component of a successful safety program, and it must be demonstrated in words and action.
- *Policy and Principles.* The company safety policy and principles must be based on the company's values, mission, and vision and must be communicated effectively throughout the organization.
- *Integrated Organizational Structure.* Safety must be integrated as a core value throughout the company, and good safety practices must be enforced by supervisors at all levels.
- *Line Management Accountability and Responsibility.* Company leaders are held responsible and accountable for the safety culture within the company and the safe performance of subordinates.
- *Goals, Objectives, and Plans.* Safety management systems require continued evaluation, identification of challenging goals and objectives, and adoption of a continuous improvement process.
- *Safety Personnel.* Experienced and knowledgeable safety professionals are needed to provide support to supervisors regarding regulatory and technical issues related to good safety management.
- *Procedures and Performance Standards.* High operational standards and aggressive goals are needed to motivate all employees to excel in good safety performance.
- *Training and Development.* A comprehensive training program consisting of orientation, initial training, and refresher training is required to ensure that all company employees have the knowledge to perform their responsibilities safely.

[1] Charles Soczek, *Implementation of Process Safety Management (PSM) in Capital Projects*, E. I. DuPont de Nemours and Company, Wilmington, DE, 2011.

- *Effective Communication.* Safety information needs to be continuously communicated within the company, on the project sites, and among the external community to improve safety performance.
- *Motivation and Awareness.* Implement incentive programs to recognize employees who make significant contributions to the company's safety performance.
- *Audits and Observations.* Conduct comprehensive safety audits at all levels within the company to monitor safety awareness and performance.
- *Incident Investigations.* Investigate any accidents, near-misses, or other incidents that could have resulted in an injury.

1.3 COST OF ACCIDENTS

Accidents and the corresponding damage that they cause to employees, property, and equipment can have a significant adverse impact on the financial condition of a construction company. It is difficult to measure precisely the cost of accidents, but they have a significant adverse impact on employee well-being, productivity, and morale. Direct costs of accidents are those costs covered by insurance, and the cost of the insurance coverage is a function of the insurance underwriter's assessment of the risk posed and the construction company's claims history. Indirect costs are all other costs not recovered thorough insurance coverage. The indirect costs associated with an accident often are up to four times the direct costs.

The direct costs of accidents include:

- Workers' compensation insurance premium cost
- Liability insurance premium cost
- Equipment liability insurance premium cost
- Legal expenses associated with claim resolution

Workers' compensation insurance is a no-fault insurance that compensates an injured worker for the cost of medical expenses, provides supplemental income if the worker is unable to work, and provides retraining if the worker cannot perform the duties of his or her job. Insurance premiums are based on the risk presented by the craft of the worker, such as roofer, carpenter, or steelworker, and the claims record of the employer. This type of insurance will be discussed in the next chapter.

General liability insurance and equipment liability insurance cover the cost of property damage or personal injury incurred by someone not involved in the project. Premium costs for these insurance policies also are affected by claims history; the more frequent the claims, the higher the policy premium. Depending upon the type of liability insurance policies obtained, legal costs may or may not be included in policy coverage. If not covered by the insurance policy, any legal expenses would be borne by the construction company.

The indirect costs of accidents include:

- First-aid expenses
- Damage or destruction of materials

- Clean-up and repair cost
- Idle construction machinery cost
- Unproductive labor time
- Construction schedule delays
- Loss of trained manpower
- Work slowdown
- Administrative and legal expenses
- Lowered employee morale
- Third-party lawsuits

The indirect cost of an accident generally greatly exceeds the direct cost incurred. In the event of an accident, all project operations typically cease while an investigation is undertaken. Workers on the site are being paid, but little productive work is being performed, leading to unproductive labor and equipment time, which may adversely affect the project budget and construction schedule. Any damaged materials or equipment will need to be replaced, and the project site may need to be cleaned. If work on a project is stopped because a contractor's employee or a subcontractor's employee has an accident, the cost of delaying the project is assumed by the contractor and not the project owner.

1.4 CAUSES OF ACCIDENTS

Why do accidents occur on construction project sites? They may result from an unsafe act by a worker or from unsafe job conditions or both. Research into why construction accidents occur has shown that about 90% of accidents on construction sites are due to unsafe behavior and about 10% are due to unsafe job site conditions. Unsafe behavior may result from a worker's state of mind, fatigue, stress, or physical condition. This may involve attempting to do more than he or she is capable of doing, such as picking up a heavy load; engaging in unsafe work activity; or improperly responding to an unsafe situation. Overexertion is a major cause of accidents, because tired workers often are not mentally alert. A major concern in the construction industry today is the aging work force and the greater susceptibility of older workers to job-related injuries.

Some examples of causes of accidents are:

- A person detects a hazardous condition but does nothing to correct it, and an accident may result. An example may be the use of defective equipment, such as a ladder.
- A person disregards a safety policy or procedure, and an accident may result. For example, a worker not wearing gloves may get a sliver when handling lumber.
- An individual may lack proper training in how to perform a specific construction task safely and may undertake performing the work in an unsafe manner.
- An individual may misjudge the risks associated with a specific task and mistakenly choose to perform the task in an unsafe manner.

The following are types of accidents that occur on construction projects in the United States each year:

- A worker is connecting steel structural members on the fourth floor of a commercial building project and falls to the ground.
- A worker is struck in the head by a load being moved by a tower crane.
- A worker is working on a platform that collapses.
- A worker installing a pipe in an open trench is crushed when the sides of the trench collapse.
- A worker installing roofing material slips and falls to the ground.
- An electrical worker installing a circuit breaker is electrocuted.
- A brick mason working on a scaffold falls to the ground.

1.5 ROLES AND RESPONSIBILITIES

The effectiveness of a company safety and health program is directly related to management's commitment to safety. Company leaders must establish the safety culture within their companies by emphasizing the importance of safety in meetings and their visits to project sites. They must also ensure that sufficient resources are provided to support a comprehensive companywide safety program that mandates the development of a specific written injury and illness prevention plan for each project site. A company safety and health program should contain the following elements:

- Hazard analysis—assessment of the hazards
- Hazard prevention—actions to be taken to keep workers safe
- Policies and procedures for working safely—rules to be followed by employees and subcontractors
- Employee training—type and frequency of training
- Continual workplace inspection—walk-around inspections of job sites
- Enforcement of company safety policies and procedures—steps to be taken when violations occur

OSHA requires that at least one person on each job site be designated as the *competent person* who is responsible for regular inspections of the site for conformance with required safety practices and procedures. To be considered competent, the individual must be knowledgeable of the various types of work to be performed on the job site as well as all required company and legally mandated safety and health practices and procedures. A competent person may have other duties, but he or she is assigned specific safety enforcement responsibilities for the project site.

Everyone on a project—from senior management to the newest employee—has responsibility for safety:

- Project managers and superintendents are responsible for establishing and enforcing safety policies and procedures, providing necessary resources, and effectively communicating safety and health information to all people working on the site, both the contractor's employees and the subcontractors.

- Field supervisors implement and enforce those safety policies and procedures as well as conduct hazard analysis, employee training sessions, accident investigations, and safety inspections.
- Employees are responsible for following established safety procedures, reporting safety hazards, and participating in safety training and meetings.

Project leaders and field supervisors must set the standard regarding safety on their projects and enforce safety standards at all times. A continual safety awareness campaign that is focused on reducing accidents is necessary. Frequent (at least weekly) job site safety inspections should be conducted to identify hazards and ensure compliance with job-specific safety rules. Every project meeting should address safety. Foremen should conduct daily safety meetings to review the safety aspect of the tasks to be performed that day. They play a critical role in establishing and maintaining a safe job site. Safety is an everyday, hands-on responsibility of all craft labor supervisors on the project.

Supervisors must conduct training to ensure that their workers are knowledgeable of:

- The construction company's safety policies and procedures
- Specific accident prevention plans developed for the project
- Housekeeping procedures to be followed on the project
- Emergency procedures for the project site
- The proper use of all equipment to be employed on the project
- The identification of any hazardous materials to be used on the project and proper procedures for handling them

Safe construction procedures and techniques should be identified for each phase of the work to minimize the potential for accidents. Some of the ways to reduce the risk created by a hazard are:

- Modify construction techniques to eliminate or minimize the hazard.
- Guard the hazard, for example, by fencing in the site.
- Provide a warning, such as back-up alarms on mobile equipment or warning signs.
- Provide special training.
- Equip workers with PPE, such as hard hats and hearing protection.

The hazards associated with each phase of work and selected mitigation strategies should be discussed with both contractor and subcontractor work crews prior to allowing them to start work. Many superintendents require daily safety meetings prior to allowing the workers to start work. These meetings address risks and mitigation strategies for the work to be performed that day. The meetings must address general housekeeping policies, emergency procedures, proper use of equipment, as well as any hazardous materials present and proper handling procedures for these materials.

In the event an accident does occur on the project site, a thorough investigation is needed to determine the cause of the accident. This information is needed in order to devise procedures to minimize the potential of a future reoccurrence. Supervisors generally are required to complete an accident report describing the results of their investigations. Most construction companies have standard formats for accident report, but in general they require answering the following questions:

- What was the injured worker doing at the time of the accident?
- What were the job site conditions?
- What equipment was being used?
- Was the injured worker trained on the proper use of the equipment?
- Was the work being performed in accordance with company safety policies and procedures?
- Was the injured worker wearing all required PPE?
- What were the primary causes of the accident?

Many construction companies employ dedicated safety professionals who develop company safety policies and procedures, conduct training programs, and inspect project sites. These individuals, however, are technical advisers to the company leaders, who bear the ultimate responsibility for creating and maintaining an effective safety program. A company safety and health manager would have staff responsibility for establishing the company's overall safety and health program. While the safety and health manager develops the program, company supervisors are responsible for its implementation. In addition, the company safety and health manager typically performs the following tasks:

- Conducting safety audits of construction sites and project offices
- Inspecting construction operations
- Conducting safety and health training
- Analyzing potential hazards
- Conducting accident investigations

1.6 ACCIDENT PREVENTION

Accident prevention requires a commitment to safety, proper equipment and construction procedures, regular and knowledgeable project site inspection, and good planning. Everyone working on the job site must understand the need to comply with all safety procedures and policies and be alert for any unsafe conditions.

Accident prevention requires an ongoing program that involves everyone working on a project work site. Techniques that superintendents should use include:

- Involving everyone in the identification of potential hazards
- Involving everyone in the development of procedures to eliminate or mitigate the hazards identified
- Teaching workers on the proper use of PPE and ensuring that the equipment is properly worn at all times

- Teaching workers safe working practices and ensuring that they do not overexert themselves
- Teaching workers the importance of good housekeeping practices and ensuring that the job site is kept orderly and clean
- Establishing emergency procedures for the project site and ensuring that all individuals working on the site understand what to do in emergency situations

The superintendent must consider whether a hazard can be engineered to a safe state (for example, using guardrails around floor openings or using trench boxes in open trenches). The next step is to create administrative controls such as policies, procedures, rules, and training sessions that do not present a physical barrier to a hazard but rely on good decision making. The final step is to enforce compliance with job site safety policies. The construction worker in Figure 1.2 is wearing protective equipment to minimize the potential for sustaining an injury.

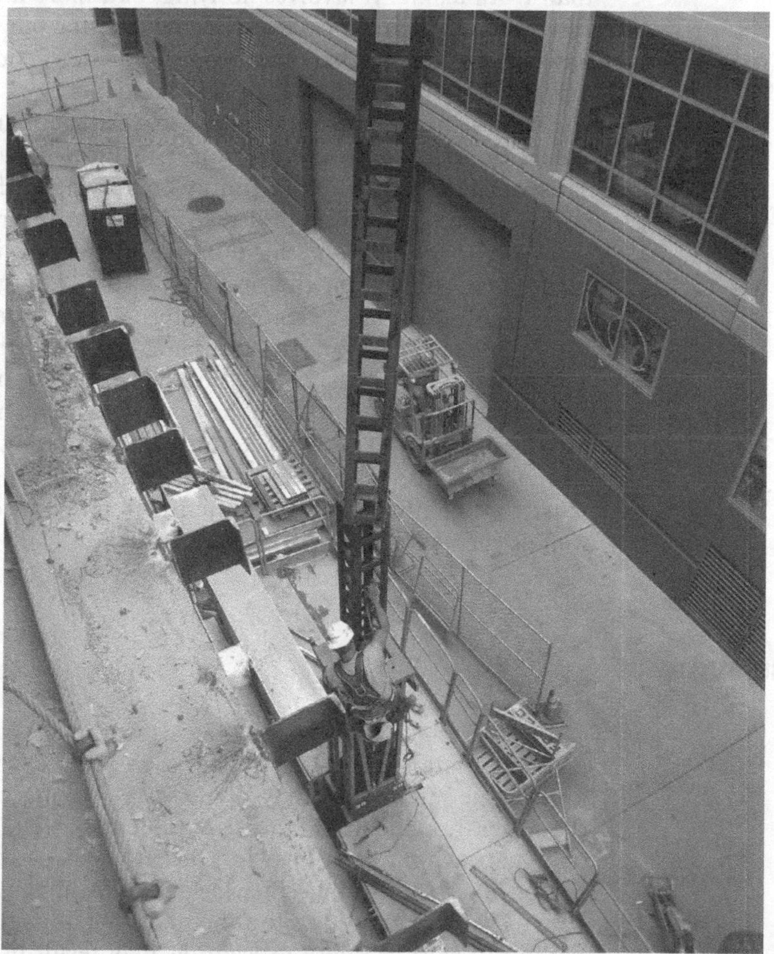

Figure 1.2 Construction worker wearing protective equipment while working above the ground

1.7 ETHICS AND SAFETY

A construction company is judged by the integrity that it demonstrates in conducting its business and how it treats its employees. A key component of a company's reputation is the ethics of the company leaders and the company's employees. This is particularly true with respect to construction safety. Ethics are the moral standards used by people in making business and personal decisions. The moral standards that we use to guide our decision making make up our ethics, which helps us to decide what is right and what is wrong.

Ethics involves determining what is right in a given situation and then having the courage to do what is right. This means taking action if an unsafe act is observed. Each decision has consequences, often to us as well as to others. Generally there are three primary ethical directives: loyalty, honesty, and responsibility. Loyalty may be requested from many groups and institutions. Honesty is more than truth telling. It involves no lying, but, more importantly, involves the correct representation of ourselves, our actions, and our views. Responsibility means anticipating the potential consequences of our actions and taking measures to prevent harmful occurrences such as accidents.

Construction company employees will face many ethical dilemmas in the execution of a construction project. Ethical behavior is a difficult area because it involves more than simply complying with legal requirements. It involves treating everyone in a responsible and fair manner. Providing a safe work environment is an ethical matter demonstrating concern for the welfare of anyone working on the project site or passing by the site. Construction company leaders must create an internal environment that promotes and rewards ethical behavior and set the example by their own personal actions. This is true in many aspects of company operations and is critical in creating a work safely culture within the company. If a company leader observes an unsafe activity and takes no action to correct the situation, he or she demonstrates a lack of interest in safety. Therefore, construction company leaders must continually demonstrate their commitment to safety by their actions as well as their statements.

1.8 SUMMARY

Construction sites are very dangerous workplaces, and numerous disability injuries and fatalities occur on them each year. There are two major aspects of project site safety: (1) safety of persons working on the site and (2) safety of the general public near the site. Accidents have a significant negative impact on construction operations. To prevent accidents, construction companies are responsible for providing safe working environments and ensuring that all individuals working on a job site adhere to company safety policies and procedures. Failure to enforce safety procedures may result in citations as well as lead to worker injuries.

Accidents are costly to contractors as well as adversely impact worker morale and productivity. They generally result from an unsafe act by a worker or from an unsafe job site condition or both. An individual may lack proper training to perform a task safely or may misjudge the risks associated with a specific task.

Company leaders must set the example regarding the company's commitment to safety. They must emphasize safety in their discussions with employees and provide adequate resources to support a company safety program. Accident prevention requires a commitment to safety, proper equipment and construction procedures, regular and knowledgeable project site inspection, and good planning. Providing a safe working environment is an ethical matter demonstrating concern for the well-being of people working on the project site.

1.9 REVIEW QUESTIONS

1. What are two major aspects of construction project site safety that a superintendent must consider?
2. What are the primary types of injury that occur on construction project sites?
3. What aspects of construction projects present the greatest safety challenges to project superintendents?
4. What are the primary factors that motivate safe practices on a construction project site?
5. What are the primary direct costs of an accident?
6. What are the primary indirect costs of an accident?
7. How are a contractor's general liability insurance premiums affected by the safety record of the company?
8. What are the primary causes of accidents on construction projects?
9. What are the primary responsibilities of company leaders with respect to creating a good safety culture within the company?
10. What is a competent person and what are his or her responsibilities on a project site?
11. What are the safety responsibilities of a superintendent on a construction project site?
12. What are the safety responsibilities of field supervisors on a construction project site?
13. What are the primary responsibilities of safety professionals employed by a construction company?
14. What techniques should superintendents use to prevent accidents on their job sites?
15. Why is implementing a good safety program considered an ethical responsibility of a construction company?

1.10 EXERCISES

1. Research the number of construction accidents and injuries that occurred in the United States last year. How many occurred? What were the primary types of injuries sustained by construction workers?
2. Find a construction company in your community and interview a project superintendent. What are his or her primary safety challenges? What strategies has he or she adopted to handle these challenges?
3. Find a construction company in your community that employs a safety professional. Interview the individual to learn what his or her job is and how he or she accomplishes his or her assigned responsibilities.

2

WORKER COMPENSATION AND INJURED WORKER MANAGEMENT

2.1 INTRODUCTION

The concept of workers' compensation was developed to eliminate the need for workers to go to court to obtain compensation for injuries incurred in the workplace. Proving that an accident was the result of employer negligence often was costly and time consuming. Today, all states have enacted workers' compensation statutes requiring employers to provide no-fault insurance to their employees. These laws require employees to give up the right to sue their employers for compensation resulting from injury or illness sustained in the workplace and employers to provide compensation irrespective of whether or not the employee's negligence contributed to the injury. The underlying rationale for developing workers' compensation was fairness to the employees and reduction of the costs to employers as a result of workplace injuries. It is a no-fault approach to resolving workplace accidents by providing needed medical care and replacement income as well as rehabilitating injured workers.

As was stated above, the objective of workers' compensation is to minimize the need for legal action to resolve issues associated with workplace accidents. Workers' compensation laws vary from state to state, but in general, they all contain the following provisions:

- *Income Replacement.* Injured employees lose income if they are unable to work. Thus, workers' compensation provides replacement income until the injured worker is able to return to work.
- *Rehabilitation.* Medical costs for an injured worker are covered until he or she recovers from injuries sustained during the accident. In the event the employee sustained a permanent disability and is unable to return to work in his or her original work classification, vocational training is provided to enable the injured worker to return to work in a different work classification.
- *Accident Prevention.* Employer's cost of workers' compensation coverage is established in such a manner as to provide an incentive for investment in accident prevention.

- *Cost Allocation.* The cost of accidents should be allocated based on the risk of injury and based on the claims history and hazards presented by the work classification. For example, the hazards encountered by a roofer are greater than those experienced by a plumber.
- Workers' compensation provides benefits to workers who are injured on the job or who contract a work-related illness. Benefits include medical treatment for work-related injuries or illness. Temporary total disability benefits are paid while the worker recuperates away from work. If the worker is unable to return to work in his or her work classification, rehabilitation training in another work classification is provided. If the worker's condition has lasting consequences after the worker heals, permanent disability payments may be paid. In the case of a fatality, the worker's dependents receive survivor benefits.
- Workers' compensation laws were enacted by the federal government and the various states to:
  - Provide income and medical benefits to injured workers or provide income to injured workers' dependents, regardless of fault
  - Eliminate the need for legal proceedings to resolve claims associated with workplace injuries
  - Promote employer interest in providing safe working environments and establishing safe working processes and procedures
  - Prevent accidents by encouraging implementation of good workplace safety and health procedures

2.2 OVERVIEW OF WORKERS' COMPENSATION LAWS

All 50 states, the District of Columbia, Guam, and Puerto Rico have enacted statutes establishing workers' compensation programs. A separate program covers federal civilian workers. Workers' compensation programs vary across the states in terms of who is allowed to provide insurance, which injuries and illnesses are compensable, and the level of benefits. Generally, state laws require employers to obtain insurance or prove that they have the financial ability to carry their own risk and self-insure. Some states have state funds that provide workers' compensation insurance. These state funds may be monopolistic or may be competitive. Private insurance coverage can be purchased in states that do not have monopolistic state funds. The premiums paid by employers are based on the occupational classifications of their workers and their record of injury frequency and benefit payments (experience based).

Private insurance carriers are the primary sources of workers' compensation insurance. They are allowed to sell insurance in all but five states that have monopolistic state funds—Ohio, North Dakota, Washington, West Virginia, and Wyoming.[1] In these five states, employers must purchase workers' compensation insurance from a state fund, unless they are allowed to self-insure.

[1] *Workers' Compensation: Benefits, Coverage, and Costs*, August 2007, National Academy of Social Insurance, Washington, DC.

Table 2.1 2012 Permanent Partial Disability Award Schedule

Loss Sustained	Disability Award ($)
Loss of one eye	45,707
Loss of hearing in both ears	91,415
Loss of hearing in one ear	15,236
Loss of thumb	41,137
Loss of index finger	20,568
Loss of middle finger	20,568
Loss of ring finger	10,284
Loss of little finger	5,142

Source: Washington Department of Labor and Industries.

States have different policies regarding how they pay for permanent disabilities. Some pay benefits for life or to retirement age, while others limit benefits for a set period or to a specified dollar amount. Many states have developed a schedule for compensating permanent impairments based on body parts, such as loss of hearing. An example partial award schedule for the State of Washington is shown in Table 2.1. For conditions not addressed in the schedules, benefits generally are determined based on the amount of impairment or the inability of the worker to compete for a position with comparable earnings.

2.3 WORKERS' COMPENSATION INSURANCE

Workers' compensation insurance is a no-fault insurance that is mandated by state law, in that the employer cannot deny a claim by an insured employee on the basis that the employee was negligent in causing the injury. Likewise, the employee cannot sue the employer on the basis that the employer's negligence contributed to the work-related injury or disease. While an employee is unable to sue his or her employer, an employee of a subcontractor or supplier may be able to sue the general contractor, who is not his or her employer, if the general contractor allowed unsafe conditions to exist on the job site. These are known as third-party lawsuits.

The dollar value of the construction firm's liability is usually established by statute. The benefits provided generally are medical benefits, loss of earnings, and retraining if the employee is unable to perform the duties of the current position. As stated in the previous section, workers' compensation insurance is offered by monopolistic state agencies in some states, while in other states, it is available from insurance companies or carriers. Construction firms that meet state requirements may choose to be self-insuring. The essence of a workers' compensation insurance contract is that the insurer agrees, for a price (premium), to assume the liability imposed on the insured construction company by the workers' compensation statute of the state named in the policy.

Only injuries incurred while undertaking work prescribed by the employee's job description, assigned by a supervisor, or normally performed in the course of performing the work classification are covered. Falling off a ladder at the employee's home would not be covered because it did not occur as a result of his or her employment.

Injuries that are compensable typically fall into one of the following four categories:

- *Temporary Partial Disability.* Injured worker is capable of light or part-time duties and is expected to recover fully. For example, a broken toe that can allow for office work but not manual labor.
- *Temporary Total Disability.* Injured worker is incapable of any work for a period of time but is expected to recover fully with no permanent disability. This is the most common type of injury. An example would be a broken bone that requires surgery.
- *Permanent Partial Disability.* Injured worker is not expected to recover fully but will be able to work again. An example would be the loss of a finger.
- *Permanent Total Disability.* Injured worker is not expected to recover and is unable to work in any job classification.

An injured worker must be referred to a physician to determine the extent of injuries, prescribe appropriate medical treatment, determine temporary work restrictions, and establish a timeline for recovery.

The benefits paid for each type of injury vary among the states. Medical costs would be covered for each type of injury. Workers sustaining a temporary partial disability typically are assigned duties that they are capable of performing. Workers sustaining a temporary total disability often are paid two-thirds of their wages during the period of disability. Workers sustaining a permanent partial disability may receive benefits based on a standard schedule adopted by the state or, if not a schedule disability, based on the amount of disability. For example, a worker with a 25% disability may receive 25% of his or her wages. Schedule disabilities typically result from the loss of a body part, such as an arm, eye, or finger. If the worker is permanently disabled, typical benefits are two-thirds of their wages for life, until retirement, or for a set period. In the event of a worker fatality on the job site, death benefits are paid to the spouse and minor children.

In addition to the benefits discussed above, workers' compensation insurance may also fund medical and/or vocational rehabilitation. Medical rehabilitation means providing whatever treatment is needed to restore, to the extent possible, any lost ability to function normally. Vocational rehabilitation means providing education and training needed to enable the injured worker to find employment in another work classification.

Under the workers' compensation statutes of the various states, the employer's liability is limited to a specific benefit level when an employee is injured or killed on the job. The benefits are automatic and cannot be appealed by the employer. Because workers' compensation insurance is a type of no-fault insurance, a construction company cannot be sued by its employees or their heirs for additional compensation as a result of injury or death.

2.4 RELATIONSHIP TO LABOR COSTS

The costs associated with workers' compensation are included in a construction company's overhead costs. Premiums are paid by the employer and are based on the construction company's payroll and differ for each employee work classification based on the degree of injury risk being assumed by a specific craft. For example, the premium rate for a steel worker is typically higher than that for a plumber. The actual rates are determined annually by a state government agency based on past losses attributed to each work classification. This is the principle of cost allocation as discussed in the first section of this chapter. Table 2.2 illustrates the variation in base rates for various construction classifications for Washington State. The rates are determined based on the number of worker hours that were used in that work classification during the previous year and the frequency and severity of claims that resulted.

Workers' compensation premiums have two components. The first is the base rate, which is applied to each \$100 of direct employee compensation. Thus the base premium for each work classification per pay period is

$$\text{Base premium} = \frac{\text{(direct wages) (base rate)}}{\$100}$$

Table 2.2 2012 Base Rates for Selected Construction Classifications

Work Classification	Base Rates ($ per Hour worked)
Commercial concrete construction	2.39
Excavation and grading	3.18
Masonry concrete	4.75
Plumbing	2.25
Insulation installation	3.32
Roofing	6.72
Exterior painting	3.70
Wood frame construction	4.11

Source: Washington Department of Labor and Industries.

Some agencies, such as the Washington Department of Labor and Industries, use an alternative method for determining the base premium. Instead of basing the rate on employee compensation, they base their rates on the actual hours worked by employees in each work classification.

The second component is the premium modifier, known as the experience modification ratio (EMR), which is based on the company's claim history in the oldest three of the past four years. Companies starting in business without a claims history are assigned an EMR of 1.0. A company's EMR is determined as follows:

$$\text{EMR} = \frac{\text{aggregate company claims cost over 3-yr period}}{\text{average claims cost for all companies over 3 yr}}$$

The average cost of claims is determined for all companies within the state doing business in the same industrial classification. For example, if the average cost of claims is \$50,000 and the company had \$40,000 in claims during the three-year period, the company's EMR would be

$$\frac{\$40{,}000}{\$50{,}000} = 0.8$$

Companies with good safety records typically have EMRs below 1.0, while the rates for companies with many claims are generally above 1.0. The actual premium rates paid by a company are

$$\text{Modified premiums} = (\text{base premium})(\text{EMR})$$

The use of the EMR makes the premiums experienced based and provides an incentive for investment in accident prevention. For example, if the monthly base premiums for a company were \$80,000 and its EMR was 0.7, its actual premium cost would be (0.7)(\$80,000), or \$56,000. But if the company's EMR was 1.3, its actual premium cost would be (1.3)(\$80,000), or \$104,000. Thus a company's EMR has a significant effect on the company's overhead cost. Investing in safety to reduce the potential for accidents will reduce the company's EMR and result in lower workers' compensation premiums.

Workers' compensation insurance may also be sold on a retrospective basis. The state fund or the insurance carrier determines an annual premium for the construction company for a 12-month period. Near the end of the year, the state fund or the carrier examines the company's claims history during the period. If the claims costs exceed the premium paid, the construction company is required to pay an additional premium. If the claims costs are lower than the premium, the state fund or the insurance company refunds the difference to the construction company.

2.5 COST REDUCTION STRATEGIES

Because workers' compensation premiums are a significant business expense, construction companies need to adopt strategies for minimizing their cost. The following strategies are suggested as methods for reducing workers' compensation costs:

- *Have a good accident prevention program.* The most effective strategy for reducing workers' compensation costs is to minimize the potential for accidents occurring. Having a site-specific safety plan to identify all project hazards and strategies for mitigating those hazards provides a framework for accident prevention. Company leaders must emphasize safety and ensure that field supervisors do not tolerate unsafe behavior.
- *Stay in contact with all injured employees.* Injured workers need to know that their employer is concerned about their well-being. They should be encouraged to return to work, even in a limited manner, as soon as they are medically able.
- *Have a good return-to-work program and use it.* The sooner an injured worker returns to work, the lower the claim will be. Ensure that the worker's physician agrees that he or she can return to work and determine what work limitations the physician stipulates.
- *Conduct thorough investigations of any accidents.* The key to reducing workers' compensation costs is to eliminate accidents. Any accident that occurs needs to be fully investigated to determine its cause. Measures then need to be implemented to avoid reoccurrence of such an accident.

2.6 INJURED WORKER MANAGEMENT

Injured workers need to be returned to the work force as soon as possible. If a worker is injured on a project, the first step is to take the injured worker to the nearest medical clinic to have the worker checked by a doctor. The doctor will determine the extent of injuries and the type of work the injured worker is able to perform. Based on the doctor's instructions, the construction company should devise a return-to-work strategy for the worker. This may involve recuperation, physical therapy, shorter work hours, and/or alternative work assignments. Keeping the worker connected with the workplace is both good medicine for the worker and good business for the company. Research has shown that effective return-to-work strategies promote faster recovery and prevent a downward spiral into disability.

Providing return-to-work options benefits the injured worker and reduces the financial impact on the company's workers' compensation premiums. In addition to reducing claim costs and insurance premiums, an effective return-to-work strategy:

- Encourages communication between the employer and the injured employee—a key factor in the employee's recovery
- Allows an experienced employee to continue working for the company

- Keeps the loss of productivity to a minimum
- Reduces the cost of training replacement employees
- Keeps the injured employee active and speeds medical recovery
- May reduce the risk of reinjury
- Provides a sense of job security to the injured employee
- Allows the injured employee to maintain contact with co-workers
- Shows that the company values the injured employee and his or her contributions to the company

The employer may offer a transitional job to an injured employee to enable him or her to return to work when restrictions preclude performing the job held when the injury occurred. This reconnects the injured employee with the construction company and demonstrates the employer's concern for the welfare of the employee. There are three types of transitional jobs that may be offered:

- *Modified work* involves an adjustment or alteration to the way in which a job is normally performed in order to accommodate the employee's physical restrictions.
- *Part-time work* involves working less than a normal work schedule because the doctor has not released the injured worker for full-time work.
- *Alternative work* is a different job within the company that meets the physical restrictions the doctor prescribed. For example, an injured carpenter may be employed in the estimating department until his or her physician authorizes return to field work.

Individuals who sustained an injury that results in permanent disability need to be given vocational training in a work classification for which they are medically qualified. This enables the injured employee to return to the work force in a different job classification. An injured worker's physician needs to verify that he or she is physically qualified for the new work classification before being allowed to start the vocational training.

2.7 SUMMARY

Workers' compensation was developed to eliminate the need for workers to go to court to obtain compensation for injuries sustained in the workplace. It is a no-fault approach in which employees give up the right to sue their employers and employers provide compensation irrespective of who caused the accident. Workers' compensation has four main objectives: income replacement, rehabilitation, accident prevention, and cost allocation.

All 50 states have established workers' compensation programs, but there is considerable variation among state programs. Some states have established state funds while others rely on private insurance companies. Some states also allow contractors to be self-insuring. Injuries that are compensable are temporary partial disability, temporary total disability, permanent partial disability, and permanent total disability. The benefits paid for each type of injury vary by state. Medical costs and replacement income are paid to injured workers. Medical and vocational rehabilitation may also be provided.

Workers' compensation insurance premiums are based on a construction company's payroll and differ for each worker classification. The rates are then modified by the experience modifier, which is based on the company's claim history in the oldest three of the past four years. Because workers' compensation is a significant business expense, construction companies need to have a good accident prevention plan, stay in contact with all injured employees, have a good return-to-work program, and thoroughly investigate all accidents. Any injured workers need to be returned to the work force as soon as possible to control claims costs. As soon as authorized by their physicians, injured workers need to be returned to work even if it is in another work classification.

2.8 REVIEW QUESTIONS

1. Why were workers' compensation laws established by the individual states?
2. What are the four major provisions of workers' compensation statutes?
3. What are three types of workers' compensation insurance?
4. How is the dollar value of the construction company's workers' compensation liability determined?
5. What are monopolistic state agencies?
6. What are four types of injuries that are compensable in workers' compensation insurance?
7. How are benefits determined?
8. Why are the base rates for roofers higher than those for plumbers?
9. What is the experience modification ratio, and how is it determined?
10. What are four strategies for reducing workers' compensation costs?

2.9 EXERCISES

1. Contact a construction company in your area and determine what program they have for managing injured employees.
2. Review the workers' compensation statute for your state. What specific benefits does it provide for injured workers?
3. Determine what restrictions are placed on vocational rehabilitation in the workers' compensation statute for your state.

3

ACCIDENT PREVENTION PROGRAM

3.1 INTRODUCTION

A written safety and health program, also known as the Occupational Health and Safety Management System, Injury and Illness Prevention Program, and Accident Prevention Program (APP), helps set out standard safety policies and procedures so that they can be understood and communicated consistently among all stakeholders. The program should prescribe: (1) the authority, responsibility, and accountability of all parties concerned, (2) measures for systematic identification and control of hazards on the job site, and (3) required initial and refresher training. Although federal regulations do not require employers to have written safety and health programs, many state-based safety and health standards, such as the ones from Washington State, specifically require the development of written and comprehensive APPs by employers. The Washington Administration Code (WAC) 296–800–140 not only asks employers to develop written APPs but also requires the APPs to be effective in practice. Therefore, any shortcut, such as simply copying a program from another project/organization, is accident prone and does not address project-specific conditions.

An effective safety and health program must be organization and site specific in order to account for all potential project hazards. It must prescribe policies, practices, and procedures that are performance oriented. Besides the practical implication and legal requirement, safety and health programs also serve as valuable means for injury prevention. Previous research, such as the survey conducted by the Nebraska Safety Council in 1981, concluded that companies with no written safety and health programs are likely to have more accidents than those with written programs. A more recent study, "Financial decision-makers' views on safety: What SH&E professionals should know," published in *Professional Safety* in 2009, positively confirmed the impact of APPs. For instance, California began to require an injury and illness prevention program in 1991. Five years after this requirement was put in place, California had a net decrease in injuries and illnesses of 19%.

Federal guidelines suggest that an effective safety and health program cover:

- Management commitment and employee involvement
- Work site analysis
- Hazard prevention and control
- Training/education

A discussed in Chapter 1, a company safety and health program should contain the following elements:

- Hazard analysis
- Hazard prevention
- Policies and procedures for working safely
- Employee training
- Continual workplace inspection
- Enforcement of company safety policies and procedures

This chapter introduces the common elements of a written safety and health program that complies with federal guideline suggestions and also presents these elements in sections that make sense for the everyday construction operation. Strategies to create a site-specific safety and health program are also discussed throughout the chapter. An example APP for a fire station project in northern Seattle is available in Appendix B and is discussed in detail in Chapter 5 to show how site-specific conditions are taken into consideration during the preparation of an APP.

3.2 COMPANY COMMITMENT AND CULTURE

A company *safety and health policy* is the first and the most important element of a safety and health program. It is a statement of the construction company management's commitment to a safe and healthful work environment. The role of management in creating a positive safety attitude is to provide the needed resources so that responsible individuals have the ability to work safely and to hold all members of the company accountable for safety in all endeavors.

Although employers ultimately bear the legal responsibility for safety and health, the policy helps deliver the message that company safety and health rules will be enforced and employees are expected to work in a safe and healthy manner. The policy, besides usually being signed by a company's chief executive officer or president, can be further emphasized by articulating other corresponding safety and health program elements in order to form a systematic approach for safety and health management. An example safety and health policy for CBE Construction is shown in Figure 3.1.

CBE Construction
15478 East Marginal Way
Seattle, Washington 98101

Safety and Health Policy

It is the policy of CBE Construction to provide a safe and healthy working environment for all employees and other persons working on our project sites. We believe that accidents are preventable and that creating and maintaining a safe and healthy work environment is a team effort requiring the involvement of everyone working on our work sites. No task is so important that someone should violate a safety rule or incur a risk of injury or illness in order to accomplish it. Anyone working on one of our job sites must wear all required personal protective equipment and comply with all safety rules. Anyone observing an unsafe activity is required to notify their supervisor. Supervisors must inspect the work place daily for any hazards before allowing workers to enter the job site and initiate work. Our goal is that everyone who comes to work on our job sites goes home at the end of each day without incurring an injury.

Jack Smith
Jack Smith
President

Figure 3.1 Safety and health policy for CBE Construction

Safety and health goals help describe the company's commitment to maintaining a safe and healthy working environment in quantitative terms. These goals should be challenging but realistic. "We are going to do our best" sounds more like an excuse for errors than a goal. Having measurable targets such as "zero fatal accidents during the year," "maintain a less than 3 incidence rate," or "reduce job-related injuries by 25%" is the key. It is very important to have a goal of no fatal accidents. Zero is the only acceptable standard with respect to fatalities on job sites. Having no recordable accidents also is a good goal and is achievable with a strong safety and health program.

It is quite common to see signs on construction sites like the one in Figure 3.2 that mark the number of days since the last recordable incident. Informing employees of their job site safety and health performance as such with respect to the predefined goal helps raise employee awareness and motivation.

Roles and responsibilities must be clearly defined at all levels within the company. The role of management is to provide the resources needed for workers to work safely and to require workers to use those resources and perform their work in a safe manner. In companies with successful safety and health programs, workers feel that they have a role in job site safety and accept responsibility for

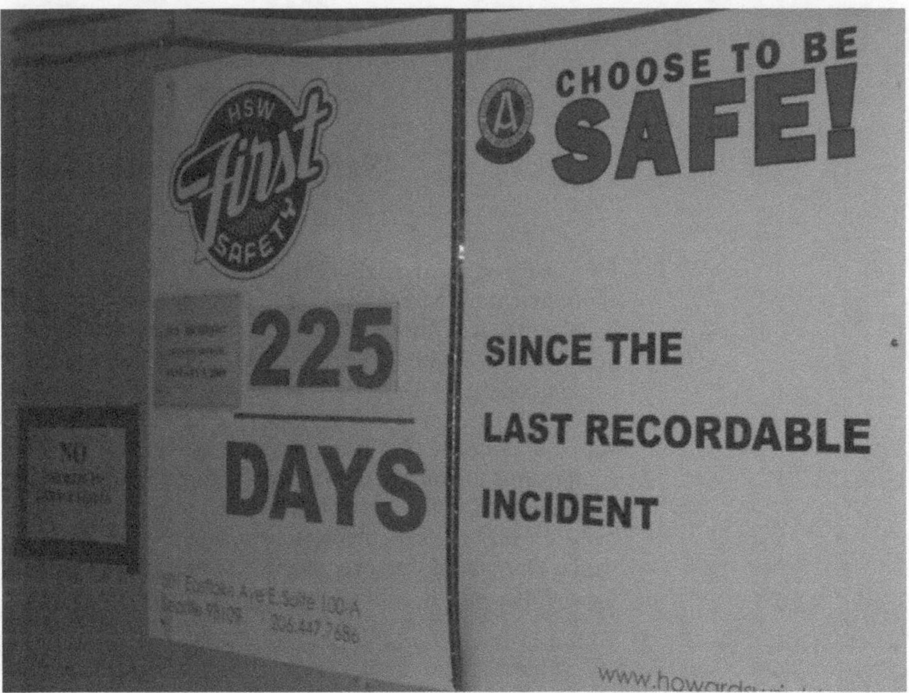

Figure 3.2 Construction job site safety performance displayed on-site

their workplace. When workers are involved in decisions that affect safety, the result is a safe project site. Workers will support a company when they know that management is committed to safety and wants to be made aware of any safety issues associated with completing the project. Everyone has a role in making the project site a safe working environment:

- Company leaders must be responsible for enforcing safety rules and providing necessary resources. They must establish the company's safety and health policies, set goals for safe performance, and set the example by their behavior.
- Supervisors need to understand and implement the administration procedures pertaining to the safety and health program. They are closest to the workers and play a key role in implementing an effective company safety and health program.
- Workers should be able to recognize hazards and apply proper hazards mitigation solutions as well as comply with all company safety and health policies.
- Safety and health professionals monitor the implementation of company safety and health policies, conduct job site inspections, and ensure that company employees receive needed training.

- Beside these individual roles, the construction company should organize a safety committee to assess company safety and health performance and to provide advice regarding initiatives that should be undertaken to improve performance. The organization and function of the company's safety committee should be stated clearly in the company's safety and health program. In some state programs (e.g., Washington State), for employers with a certain number or more employees, the regulations specifically call for the establishment of a safety committee as well as employee participation in the committee.

3.3 PROJECT ADMINISTRATION

Accountability is a key component of a company's safety and health program. It is not sufficient to just have a set of policies and procedures that must be followed. There must be consequences for failure to follow them. People working on a job site need to understand all company safety requirements and the consequences for failure to follow them. Fines may be assessed for failure to properly wear required PPE, and written warnings may be issued for performing work in an unsafe manner. Individuals who continue to fail to follow required safety procedures may be let go.

Some construction companies use immediate termination of employment as the only option for failure to follow required safety and health procedures, while others adopt a more progressive philosophy with verbal warning being the first step of discipline, followed by written warning, pay suspension, and then employment termination. For construction projects, although subcontractors maintain their own safety and health programs, the breach of contract clause allows a general contractor to remove a subcontractor's employee from its project if the employee is performing work in an unsafe manner. The potential consequences of failure to comply with company safety policies and procedures need to be described in the company safety and health program.

On the positive side, employers do provide incentives for projects that meet the predefined safety and health goals (e.g., 500 days without any injuries or accidents) or for employees that work safely. The incentives can be in various forms ranging from cash to public recognition. Nonbelievers of the incentive programs claim that the incentives cause underreporting and favoritism. However, research findings revealed effective reductions in injuries and illnesses when adequately designed incentive programs were in use.

Safety inspections, also known as field inspections, are one of the critical means to promote workplace safety and health. They are important components of a construction company's safety and health program. These inspections are a proactive approach to identifying and eliminating job hazards as well demonstrating the importance that company leaders place on workplace safety and health. Some of these inspections may be performed by project supervisors, while others may be conducted by safety and health professionals. For construction projects, a safety inspection is commonly administered through job walks, as shown in Figure 3.3.

Figure 3.3 Job walks on construction sites are an important means of safety assurance

The frequency of job walks depends on the size of a project and is a function of the project complexity and the risks associated with its construction. At a minimum, weekly safety inspections of job sites are recommended. A project-specific safety and health program should specify the job walk frequency and the parties responsible for the walk. For large or complicated projects that have on-site safety professionals, job walks take place at least once a day. For medium-sized projects, safety professionals might visit the job sites once a week. Smaller projects such as those that involve only simple renovation might only be visited by safety professionals once a month but inspected weekly by field supervisors. Checklists are often used during safety inspection and should be tailored to address site-specific conditions. While large contractors are more resourceful and might adopt innovative technologies to assist with safety inspections, small- to medium-sized contractors may still rely on paper and pencil.

Some argue that safety inspections only provide a snapshot of the workplace condition and do not reveal overall safety performance or catch near-miss errors. A near-miss event is an unplanned event that did not result in injury, illness, or damage but had the potential to do so. Only a fortunate break in the chain of events prevented an injury, fatality, or damage. Field workers often are the only ones who are aware of near misses, making such events difficult to catch through safety inspections. On the other hand, safety inspections do offer a preventive opportunity to promote workplace safety and health, before expensive accidents and injuries take place. Figures 3.4 and 3.5 provide an example construction inspection checklist from the Washington State Department of Labor and Industries (WA L&I). The checklist is available on WA L&I's website as a part of a sample APP for the construction industry.

Safety and Health Inspection Check List – Sample 1

Job site: ______________________ Date: ______________

This format is intended only as a reminder to look for unsafe practices, potential and/or near miss incidents.

(S) indicates **S**atisfactory **(U)** indicates **U**nsatisfactory

Date of inspection/walk around												
Machinery												
Point of operation guard												
Belts, pulleys, gears, shafts, etc.												
Oiling, cleaning, and adjusting												
Maintenance and oil leaks												
Pressure equipment												
Steam equipment												
Air receivers and compressors												
Gas cylinders and hoses												
Unsafe practices												
Excessive speed of vehicles												
Improper lifting												
Smoking in dangerous places												
Horseplay												
Running in aisles or on stairs												
Improper use of air hoses												
Removing machine guards												
Working under suspended loads												
Working on machines in motion												
First aid												
First aid kits												
Stretchers and fire blankets												
Emergency showers												
Eyewash stations												
All injuries and illnesses reported												
Hazard Communications												
Acids and caustics												
Solvents												
Dusts, vapors, or fumes												
Radiation												
New chemicals/processes												

J-1

Figure 3.4 Sample safety and health inspection check list (Part I)

Safety and Health Inspection Check List – continued

Job site: ______________________ Date: ______________________

(S) indicates Satisfactory **(U) indicates Unsatisfactory**

Date of inspection/walk around												
Tools												
Power tools, wiring, and grounding												
Hand tools (condition)												
Use and storage of tools												
Personal protective equipment												
Goggles or face shield												
Substantial footwear												
Hard hats												
Gloves												
Respirators												
Fall protection equipment												
Other protective clothing												
Fire protection												
Extinguishing equipment												
Exits, stairs, and signs												
Storage of flammable materials												
Material handling equipment												
Power trucks and hand trucks												
Elevators												
Cranes and hoists												
Conveyors												
Cables, ropes, chains, slings												
Housekeeping												
Aisles, stairs, and floors												
Storage and piling of materials												
Wash and locker rooms												
Light and ventilation												
Disposal of water												
Yards and parking lots												
Bulletin boards												
Only safety and health materials posted												
Neat and attractive												
Display regularly changed												
Well-illuminated												

(Customize the checklist above with any additional information.)

J-2

Figure 3.5 Sample safety and health inspection check list (Part II)

Accident investigation can be considered an after-the-fact approach designed to identify the cause of accidents and to prevent similar situations from reoccurring. All accidents should be investigated regardless of their severity because the same causes could also result in disastrous fatalities. A comprehensive investigation is needed to determine the cause of the accident and to identify measures that should be taken to prevent reoccurrence. The investigation needs to be conducted soon after an accident occurs before the site conditions are altered and witness memories begin to fade. The purpose of the investigation is to determine facts, not to find fault.

The following questions should be addressed during an accident investigation:

- What type of work was being performed when the accident occurred?
- What specific tasks were being performed by the injured persons?
- Were the injured persons properly trained to perform their assigned tasks?
- Were the tasks being performed in accordance with company safety policies and procedures?
- Was the equipment being used for the tasks in good working order and operated in a safe manner?
- What were the work site conditions when the accident occurred?
- Were the injured persons wearing proper PPE?
- Were supervisors present when the accident occurred?
- Were other workers present when the accident occurred?
- What measures could have been adopted that would have prevented the accident?

For construction projects, job site supervisors or superintendents are typically responsible for most accident investigations. Protocols for conducting an accident investigation should clearly describe the necessary steps, such as accident site isolation, photographing the site, and interviewing witnesses, to help describe the events that led up to the accident. For workers' compensation claims, investigation protocols might require employees who are involved in an accident to take postaccident drug tests. Most construction companies have a standard format for conducting an accident investigation, such as the sample shown in Figure 3.6.

Record keeping and reporting guidelines are for both business management and OSHA or state program compliance. The guidelines typically cover all property-related incidents, injuries, illnesses, fatalities, and near-miss events, although different employers have different definitions of what constitutes a near-miss situation. A project-specific safety and health program should describe the reporting and communication process with the contact information of responsible personnel listed. According to OSHA, employers that have 11 or more employees must maintain records of occupational injuries and accidents. The records must be kept for five years and be available for a compliance officer's inspection. While it is common that companies use proprietary recording keeping forms for internal reporting and communication, federal standards require that employers use OSHA Form 300 to record any injury or illness involving any work-related fatality or the in-patient hospitalization of 3 or more employees. Specific federal OSHA record keeping and reporting requirements are discussed in Chapter 4.

CBE Construction
15478 East Marginal Way
Seattle, Washington 98101

Accident Investigation Form

Project Name: ______________________________

Project Location: ______________________________

1. Name of injured person: ______________________________

2. Date and time of accident: ______________________________

3. Job title of injured person: ______________________________

4. Nature of injuries sustained by injured person: ______________________________

____

5. Describe the accident and how it occurred: ______________________________

____

____

6. What was the cause of the accident?: ______________________________

____

____

7. Were all company safety policies being followed? ________ If, no, explain. ________

____

8. Was PPE required? ________ Was all required PPE being worn? ________ If no, explain

____

9. Describe the job site conditions at the time of the accident: ______________________________

____

10. Had the injured person been properly trained to perform the assigned tasks? ________

11. Witnesses: ______________________________

Prepared by ________________ Date ________________

Figure 3.6 Sample accident investigation form

Training can be as formal as a classroom setting or as casual as a conversational meeting right before the shift of a day begins. A safety and health program should specify:

- Types of training required
- When they are required
- Who must attend the training
- Frequency for the refreshment training

Mandatory orientation training is common for employees who are new to a project and covers topics such as emergency evacuation routes, shut-off valve locations, and where to find on-site first-aid kits. When a project involves unique hazards, such as a confined space or asbestos, its site-specific safety and health program should demand specialized training for those individuals who will be working in and around the hazards. Workers must be trained to:

- Be able to identify hazards on the project site
- Understand the risks presented by the hazards
- Understand company policies to prevent injury from those hazards
- Use provided safety equipment properly

At the end of each training session, a test should be administered to verify worker achievement of the training objectives. Training documentation is of particular importance for record keeping. For this reason, most employers create standard forms to record trainee information and compile checklists of topics that should be covered.

Medical assistance and *first aid* must be available and easily accessible. Medical assistance can be on-site or from nearby third-party providers. Information such as the address, phone number, and directions to the closest off-site medical assistance provider is project specific and should be posted throughout a project's job site in case of emergency situations. A safety and health program should also define the procedures to handle injuries or illnesses that require medical assistance. For example, when an employee is injured, he or she should be accompanied by the on-site supervisor or superintendent to the nearby medical service provider.

First-aid kits are commonly placed at job trailers and within an employer's company vehicles. Designated employees and supervisors should be trained in first-aid fundamentals, although many employers also ask these personnel to be CPR certified.

An *emergency response plan* delineates critical information specific to the project for timely responses and communication in the event of unexpected emergencies. An emergency is a potentially life-threatening situation, usually occurring suddenly and unexpectedly. For construction, such emergencies can range from natural (e.g., earthquake) or human error (e.g., fire) disasters to terrorist attacks. When an emergency occurs, immediate response is essential. Speed in responding can make the difference between life and death. All individuals working on the job site need to know what to do in the event of an emergency. The plan should contain information about the job site itself for communicating with outside responders, identify job site emergency response personnel, list important phone numbers and addresses (e.g., utility and medical service providers), point out emergency evacuation routes, and explain how reporting and communication should take place (e.g., to account for all workers on-site).

Employee communication is the key to an effective safety and health program and should take a variety of forms such as using multimedia safety awareness videos. Company newsletters, safety alert notifications, safety-related meetings, fliers/posters, and trainings are all potential channels that help employers communicate with their employees.

3.4 PERSONAL PROTECTIVE EQUIPMENT

Hazards exist everywhere on a construction job site. When engineering, work practice, and administrative controls are not feasible or do not provide adequate protection, the company must provide workers with PPE and require its use. PPE is equipment worn to minimize exposure to a variety of hazards. It includes protective equipment for eyes, face, head, arms, hands, legs, and feet; respiratory devices; and protective shields. PPE is the least effective way to protect workers, because it does not eliminate or reduce the hazard. It only places a barrier between the worker and the hazard. Some examples of PPE are shown in Table 3.1.

All PPE clothing and equipment should be of safe design and construction and maintained in a clean and reliable manner. Selected PPE must fit well and be comfortable to wear. All workers need to be trained in the proper use of all required PPE. Before deciding what PPE is needed, a job site assessment is needed to determine what hazards are likely to be present. The assessment form shown in Figures 3.7, 3.8, and 3.9 can be used for this purpose.

If PPE is determined to be necessary, it should be selected to match the anticipated hazards. The PPE should be issued to all workers to be employed on the project site, and they should be instructed in its proper use and the importance of wearing the PPE at all times when working on the site. Supervisors should continuously monitor the workers to ensure proper wearing of all required PPE. The most common types of PPE fall into the following six categories:

- *Eye Protection*. Construction workers can be exposed to many hazards that pose danger to their eyes. Safety spectacles or goggles usually are required on all construction sites to minimize the risk of eye injury.
- *Foot Protection*. Construction workers may be exposed to hazards that may result in crushing from objects dropped on the foot, punctures from protrusions on the job site, or sprains from lack of ankle support. Steel-toed safety shoes or boots are generally required of all workers to minimize the potential for foot injury.

Table 3.1 Examples of PPE

Body Part	Example of PPE	Example of Hazard
Head	Hard hat	Contact from falling object
Face	Face shield	Liquid chemical splash
Eyes	Safety glasses	Impact from flying wood chips
Arms and hands	Leather gloves	Sharp or rough objects
Feet	Steel-toed boots	Contact from falling object
Potentially life threatening	Body harness	Fall from roof
Ears	Ear muffs	Loud noise from equipment
Lungs	Respirator	Vapors from paint or cleaning materials

PPE Hazard Assessment Certification Form

***Name of work place:** ______________________	***Assessment conducted by:** ______________
***Work place address:** ______________________	***Date of assessment:** ______________
Work area(s): ______________________	**Job/Task(s):** ______________
*Required for certifying the hazard assessment.	Use a separate sheet for each job/task or work area

EYES		
Work activities, such as: ☐ abrasive blasting ☐ chopping ☐ cutting ☐ drilling ☐ welding ☐ punch press operations ☐ other: ____ ☐ sanding ☐ sawing ☐ grinding ☐ hammering	Work-related exposure to: ☐ airborne dust ☐ flying particles ☐ blood splashes ☐ hazardous liquid chemicals ☐ intense light ☐ other: ____	Can hazard be eliminated without the use of PPE? Yes ☐ No ☐ If no, use: ☐ Safety glasses ☐ Safety goggles ☐ Shading/Filter (#____) ☐ Welding shield ☐ Other: ____ ☐ Side shields ☐ Dust-tight goggles
FACE		
Work activities, such as: ☐ cleaning ☐ cooking ☐ siphoning ☐ painting ☐ dip tank operations ☐ other ____ ☐ foundry work ☐ welding ☐ mixing ☐ pouring molten metal	Work-related exposure to: ☐ hazardous liquid chemicals ☐ extreme heat/cold ☐ potential irritants: ____ ☐ other: ____	Can hazard be eliminated without the use of PPE? Yes ☐ No ☐ If no, use: ☐ Face shield ☐ Shading/filter (#____) ☐ Welding shield ☐ Other: ____
HEAD		

Work activities, such as:
☐ building maintenance
☐ confined space operations
☐ construction
☐ electrical wiring
☐ walking/working under catwalks
☐ walking/working under conveyor belts
☐ walking/working under crane loads
☐ utility work
☐ other: ____
	Work-related exposure to:
	☐ beams
	☐ pipes
	☐ exposed electrical wiring or components
	☐ falling objects
	☐ machine parts
	☐ other: ____
		Can hazard be eliminated without the use of PPE?
		Yes ☐ No ☐
		If no, use:
		☐ Protective helmet
		  ☐ Type A (low voltage)
		  ☐ Type B (high voltage)
		  ☐ Type C
		  ☐ Bump cap (not ANSI-approved)
		☐ Hair net or soft cap
		☐ Other: ____

Figure 3.7 Sample PPE hazard assessment form—Part 1

- *Hand Protection*. Construction workers may be exposed to many sharp or rough objects on a job site. Consequently, most construction companies require all workers to wear protective gloves at all times when working on the site.
- *Head Protection*. Protecting construction workers from head injuries is a key component of any safety program. Wearing a safety helmet or hard hat should be required of anyone present on a construction site. Objects may fall from above or workers may bump their heads against fixed objects.
- *Hearing Protection*. Exposure to high noise levels can cause hearing loss or impairment. If high noise levels are present, construction workers should be issued earplugs or earmuffs to reduce the noise exposure to acceptable levels.
- *Respiratory Protection*. Construction workers who are working in areas where there may be oxygen deficiency or unhealthy gases or vapors need to wear properly fitted respirators.

HANDS/ARMS		
Work activities, such as: ☐ baking ☐ cooking ☐ grinding ☐ welding ☐ working with glass ☐ using computers ☐ using knives ☐ dental and health care services ☐ other: _____ ☐ material handling ☐ sanding ☐ sawing ☐ hammering	Work-related exposure to: ☐ blood ☐ irritating chemicals ☐ tools or materials that could scrape, bruise, or cut ☐ extreme heat/cold ☐ other: _____	Can hazard be eliminated without the use of PPE? Yes ☐ No ☐ If no, use: ☐ Gloves ☐ Chemical resistance ☐ Liquid/leak resistance ☐ Temperature resistance ☐ Abrasion/cut resistance ☐ Slip resistance ☐ Protective sleeves ☐ Other: _____
FEET/LEGS		
Work activities, such as: ☐ building maintenance ☐ construction ☐ demolition ☐ food processing ☐ foundry work ☐ logging ☐ plumbing ☐ trenching ☐ use of highly flammable materials ☐ welding ☐ other: _____	Work-related exposure to: ☐ explosive atmospheres ☐ explosives ☐ exposed electrical wiring or components ☐ heavy equipment ☐ slippery surfaces ☐ tools ☐ other: _____	Can hazard be eliminated without the use of PPE? Yes ☐ No ☐ If no, use: ☐ Safety shoes or boots ☐ Toe protection ☐ Electrical protection ☐ Puncture resistance ☐ Anti-slip soles ☐ Metatarsal protection ☐ Heat/cold protection ☐ Chemical resistance ☐ Leggings or chaps ☐ Foot-leg guards ☐ Other: _____
BODY/SKIN		
Work activities such as: ☐ baking or frying ☐ battery charging ☐ dip tank operations ☐ fiberglass installation ☐ irritating chemicals ☐ sawing ☐ other: _____	Work-related exposure to: ☐ chemical splashes ☐ extreme heat/cold ☐ sharp or rough edges ☐ other: _____	Can hazard be eliminated without the use of PPE? Yes ☐ No ☐ If no, use: ☐ Vest, jacket ☐ Coveralls, body suit ☐ Raingear ☐ Apron ☐ Welding leathers ☐ Abrasion/cut resistance ☐ Other: _____

Figure 3.8 Sample PPE hazard assessment form—Part 2

BODY/WHOLE		
Work activities such as: ☐ building maintenance ☐ construction ☐ logging ☐ utility work ☐ other: _____	Work-related exposure to: ☐ working from heights of 10 feet or more ☐ working near water ☐ other: _____	Can hazard be eliminated without the use of PPE? Yes ☐ No ☐ If no, use: ☐ Fall arrest/restraint: Type: _____ ☐ PFD: Type: _____ ☐ Other: _____
LUNGS/RESPIRATORY		
Work activities such as: ☐ cleaning ☐ mixing ☐ painting ☐ fiberglass installation ☐ compressed air or gas operations ☐ other: ☐ pouring ☐ sawing	Work-related exposure to: ☐ irritating dust or particulate ☐ irritating or toxic gas/vapor ☐ other:	Can hazard be eliminated without the use of PPE? Yes ☐ No ☐
EARS/HEARING		
Work activities such as: ☐ generator ☐ ventilation fans ☐ motors ☐ sanding ☐ pneumatic equipment ☐ punch or brake presses ☐ use of conveyors ☐ other: _____ ☐ grinding ☐ machining ☐ routers ☐ sawing	Work-related exposure to: ☐ loud noises ☐ loud work environment ☐ noisy machines/tools ☐ punch or brake presses ☐ other: _____	Can hazard be eliminated without the use of PPE? Yes ☐ No ☐

Figure 3.9 Sample PPE hazard assessment form—Part 3

3.5 PHASED SAFETY PLANNING

Phased safety planning is a systematic approach that examines the logical sequence of required construction activities and highlights the potential hazards behind these activities. A hazard is a condition that, if uncorrected, may lead to an injury or illness. A hazard analysis involves determining all of the tasks to be performed, the conditions under which they are to be performed, and the likelihood of an accident or injury.

The planning process requires a good understanding of construction practices and site-specific conditions and is often led by a job site's supervisor with the participation of other project stakeholders such as owner's representatives or subcontractors of major operations. For renovation projects where existing tenants might be in and out of the project sites, the coordination and communication with project owners become essential to phased safety planning so that restrictions and issues related to public safety can be discussed. Dangerous or complex operations (e.g., critical lifts) that come with severe or anticipated hazards (e.g., hoisting failure) should be specifically called out in a company's safety and health program, regardless of whether the operations are to be completed by the company's own workforce or by its subcontractors. Ultimately, good phased safety planning should produce a list of all hazardous activities that will be employed in completing the project. This list of hazardous activities will guide the identification of activity-specific hazard prevention solutions.

A good phased safety plan also influences the establishment of a project's site logistics plan because all safety-related aspects have to be considered as a part of the site logistics plan. At a minimum, a site logistics plan should specify fencing and access gate locations, required signage, circulation and access routes for material deliveries, temporary storage and lay down areas for trades on-site, emergency meet-up points, and location of trailers, toilets, wash stations, temporary utilities, dumpsters, and hoisting equipment.

The technique to identify potential hazards and propose ways to eliminate, control, or reduce the hazards during project planning is job hazard analysis (JHA). A JHA involves reviewing each task to be performed in a systematic way to determine the safest method to perform each task. It breaks a construction activity into tasks or steps, lists all potential hazards associated with each task, and suggests proper ways to eliminate or reduce the hazards to an acceptable risk level. Identifying each task to be performed and analyzing each task to identify all potential hazards associated with performing the task ensure that critical hazards are not overlooked in the JHA. This knowledge can also be used to gauge the likelihood of a hazard and the severity of its consequence in order to highlight the importance or training requirement for corresponding safety approaches.

JHA often improves not only the safety and health aspect of a project but also its productivity, management, and worker satisfaction. Table 3.2 displays an example JHA for a typical site preparation and grading activity that is under OSHA's jurisdiction. PPE like safety glasses and hard hats, which are mandatory at almost all construction job sites, are not repeated in the example JHA. Safety approaches in JHAs are often generated based on experiences, applicable safety regulations, and safety challenges such as accidents or near

Table 3.2 Example JHA for Site Preparation and Grading

Activity	Hazard	Safety Approach
Unloading equipment	Equipment sliding off trailer	Park the trailer on a firm and level surface Set equipment break while unloading Have a traffic control plan Use a flagger for traffic control Use a spotter to monitor the vicinity of personnel, buildings, and other equipment
Daily equipment walk-around	Struck by	Ensure that emergency brake is on Use three points of contacts when mounting and dismounting equipment
Site excavation and grading	Struck by	Ground personnel wear high-visibility vests Ground personnel stay outside the equipment swing radius
Site excavation and grading	Underground utility	Locate utility in advance Check for private buried lines Use a spotter when in a close proximity to utilities Hand dig when in a close proximity to utilities and use extreme cautions Wear appropriate PPEs when hand digging in a close proximity to existing utilities
Site excavation and grading	Equipment tipping over or overloading	Observe the load limit Equipment operators always wear their seat belts
Site excavation and grading	Noise	Equipment operators and nearby ground personnel wear hearing protection if the noise level is at or above 90 dBA
Loading earth/debris into dump trucks	Equipment damage	Ensure that the bucket is not overloaded
Loading earth/debris into dump trucks	Struck by	Ensure path is clear before swinging buckets
Shutting down equipment	Unexpected equipment movement	Block equipment wheels Set equipment breaks Lower attachments to the ground Lock the equipment
Public safety	Injuries to unauthorized personnel	Mark site perimeter with fences, signs, and caution tapes Required authorized personnel to wear PPEs and be escorted Ensure site security and visitor policy

misses that have occurred in the past. Therefore, learning from mistakes and accumulating corporate knowledge can improve the continuous development of JHAs. In Table 3.2, the need to use hearing protection is determined by the permissible noise exposures set forth in the OSHA standard 1926.52, Occupational Noise Exposure.

During JHA development, when multiple safety approaches are applicable, engineering controls are preferred over other means like administrative controls, PPE, or safe practices. Engineering controls eliminate or minimize hazards by physically changing the work environment to actively prevent employee exposure to the hazard. Taking concrete cutting as an example, isolating sawdust through the incorporation of enclosure devices is a demonstration of effective engineering controls. If such controls are not feasible, then manually venting the work area (i.e., safe practices), using respirators (i.e., PPE), and rotating workers so that no one is exposed to the dust for more than a certain period of time (i.e., administrative controls) can be considered as viable alternatives.

3.6 TECHNICAL SECTIONS

Technical sections represent safety standards or best practices that correspond to the hazards identified in phased project planning and provide supplemental information to JHAs. Therefore, although OSHA safety standards for the construction industry cover a wide range of safety and health topics from "personal protective and lifesaving equipment" to "cranes and derricks used in demolition and underground construction," not all topics appear in the technical sections of a company's safety and health program. Only the sections that apply to a project should be included in the project's safety and health program. If there is no confined space on-site, it does not make sense to have "confined space safety" in a safety and health program. Based on this principle, for general contractors, masonry construction and steel erection are typically subcontracted out and are rarely included in safety and health programs. Conversely, "fall protection," "scaffold safety," and "ladder safety" can be expected in almost every safety and health program in construction. The type of project being constructed also influences the focus of selected safety topics. For new construction, electrical safety would discuss the general safety procedures (e.g., the use of ground-fault-circuit interrupters, GFCI), while for jobs within existing building facilities, lock-out-tag-out becomes much more emphasized.

Below is a list of the construction-specific technical sections suggested by the WA L&I. The list is intended to serve as a starting point for construction employers and should be adjusted to reflect project-specific limitations and conditions:

- Ladder safety
- Fall protection
- Trenching and excavation
- Scaffold safety

- Motorized vehicles and equipment
- Material handling and safety
- Lock-out-tag-out
- Welding and cutting
- Hazard communication
- Respirator program
- Hearing conservation program
- Heat stress
- Confined space

Among the technical sections suggested by WA L&I, hazard communication is less understood by field professionals as it refers to the hazards of on-site chemicals and not the physical hazards that are visibly more familiar to the construction practitioners. For hazard communication, a list of all hazardous chemicals used on-site as well as the material safety data sheet (MSDS) for each of the hazardous chemicals should be available on-site and continuously updated. The goals of hazard communication are to educate companies and their employees about work hazards and how to protect themselves as well as to reduce the risk of chemical source illness or injury.

Information from MSDSs helps inform employees of ways to protect themselves as well as others when handling hazardous chemicals. Suppliers of hazardous materials must provide a MSDS for each hazardous material delivered to the site, and these sheets must be available on the job site. Each MSDS is required to contain at least the following information:

- Specific chemical identities of the hazardous chemicals involved and their common names
- Physical and chemical characteristics of the hazardous chemicals
- Known health effects and related information
- Exposure limits
- Whether the chemicals are considered to be carcinogenic
- Precautionary measures
- Emergency and first-aid procedures
- Control measures

Figures 3.10 and 3.11 contain the MSDS for acetone, a sweet-smelling chemical used in many solvents for painting applications. As the two figures illustrate, job site personnel can learn from the MSDS about acetone's permissible exposure limit, potential hazards, emergency first-aid procedures, and preventive measures that recommend chemical-handling PPEs. As of 2012, OSHA filed a new rule to align its hazard communication standards (HCS) with the Globally Harmonized System of Classification and Labeling of Chemicals (GHS). Under GHS, MSDSs will be referred to as safety data sheets (SDS) and it is expected that definition and labeling of chemical hazards will be more consistently communicated to employees after the adoption of GHS.

840

MATERIAL SAFETY DATA SHEET

Complies with OSHA Hazard Communication Standard 29 CFR 1910.1200

Date of Prep: 9/19/02

SECTION 1

[redacted] CORPORATION
[redacted] ([redacted]) [redacted]
EMERGENCY TELEPHONE ([redacted]) [redacted]

FOR INFORMATION: ([redacted]) [redacted]
- [redacted] CORPORATION
- CHEM TREC

Product Class: Ketone
Trade Name: ACETONE

Manufacturer's Code: [redacted]
NPCA HMIS: Health: 1; Flammability: 3; Reactivity: 0

Product Appearance and Odor: Clear, water-white liquid; typical, pungent odor.

SECTION 2 – HAZARDOUS INGREDIENTS

OCCUPATIONAL EXPOSURE LIMITS

INGREDIENT	CAS #	PERCENT	ACGIH TLV (TWA)	ACGIH TLV (STEL)	OSHA PEL (TWA)	OSHA PEL (STEL)	VAPOR PRESSURE
Acetone	67-64-1		500 PPM	750 PPM	750 PPM	1000 PPM	213 MM Hg @ 75° F

SECTION 3 – EMERGENCY AND FIRST AID PROCEDURES

INHALATION: Using proper respiratory protection, immediately remove the affected victim from exposure. Administer artificial respiration if breathing is stopped. If breathing is difficult, qualified medical personnel may administer oxygen and continue to monitor. Keep at rest. Get medical attention immediately.

EYE CONTACT: Flush eyes immediately with water for at least 15 minutes. Get prompt medical attention.

SKIN CONTACT: Flush with large amounts of water; use soap if available. Remove grossly contaminated clothing, including shoes, and launder before reuse. Get medical attention if skin irritation develops or persists.

INGESTION: If swallowed, do not induce vomiting. Give victim a glass of water or milk. Call a physician or poison control center immediately. Never give anything by mouth to an unconscious person.

SECTION 4 – PHYSICAL DATA

The following data represent approximate or typical values. They do not constitute product specifications.

Boiling Range: 133° (F) (I.B.P.)
Vapor Density: Heavier than air
Evaporation Rate: Slower than ether
% Volatile By Volume: 100%
Weight Per Gallon: 6.60 Lbs.
Solubility in Water: Complete

SECTION 5 – FIRE AND EXPLOSION DATA

Flammability Classification: Flammable Liquid - Class IB.

Flash Point: 0° (F) (Tag. Closed Cup).

Autoignition Temperature: 869° (F)

Lower Explosive Limit: 2.5% @ 77° (F)

Extinguishing Media: Either allow fire to burn under controlled conditions or extinguish with alcohol resistant foam and dry chemical. Try to cover liquid spills with foam.

Unusual Fire and Explosion Hazards: Extremely flammable. Vapors may cause a flash fire or ignite explosively. Vapors may travel considerable distance to a source of ignition and flash back. Prevent buildup of vapors or gases to explosive concentrations.

Special Fire Fighting Procedures: Use water spray to cool fire exposed surfaces and to protect personnel. Shut off "fuel" to fire. If a leak or spill has not ignited, use water spray to disperse the vapors.

Figure 3.10 MSDS for acetone (Part I)

Trade Name: ACETONE	Page 2 of 3

SECTION 6 – HEALTH HAZARD DATA

THRESHOLD LIMIT VALUE:	500 PPM (ACGIH-Time weighted average).
EFFECTS OF OVEREXPOSURE	
Inhalation:	Vapors or mist may cause irritation of the nose and throat. Inhilation may cause dizziness, drowsiness, loss of coordination, disorientation, headache, nausea, and vomiting. In poorly ventilated areas or confined spaces, unconsciousness and asphyxiation may result.
Skin Contact:	Prolonged or repeated skin contact may cause irritation, discomfort and dermatitis.
Eye Contact:	Severely irritating and may injure cornea if not removed promptly. High vapor concentrations are also irritating.
Ingestion:	Product may be harmful or fatal if swallowed. Material is a pulmonary aspiration hazard. Material can enter lungs and cause damage. Ingestion of this product may cause central nervous system effects, which may include dizziness, loss of balance and coordination, unconsciousness, coma and even death.
Carcinogenicity:	Acetone is not listed by the NTP, IARC, or OSHA.

SECTION 7 – REACTIVITY DATA

Stability:	Stable
Conditions to Avoid:	Heat, sparks and flame.
Incompatibility (Materials to Avoid):	Caustics, amines, alkanolamines, aldehydes, ammonia, strong oxidizing agents, and chlorinated compounds.
Hazardous Decomposition Products:	Thermal decomposition may yield carbon monoxide, carbon dioxide, adehydes and keytones.
Hazardous Polymerization:	Will not occur.

SECTION 8 – SPIL OR LEAK PROCEDURES

Steps to be taken in case material is spilled or released: Remove ignition sources, evacuate area, avoid breathing vapors or contact with liquid. Recover free liquid or stop leak if possible. Dike large spills and use absorbent material for small spills. Keep spilled material out of sewers, ditches and bodies of water. Warn occupants and shipping in surrounding and downwind areas of fire and explosion hazard and request all to stay clear.

Waste disposal method: Incinerate under safe conditions; dispose of in accordance with local, state and federal regulations.

SECTION 9 – SAFE HANDLING AND USE INFORMATION

Respiratory Protection:	Where concentrations in air may exceed occupational exposure limits, NIOSH/MSHA approved respirators may be necessary to prevent overexposure by inhalation.
Ventilation:	Exposure levels should be maintained below applicable exposure limits - see Section 2. This product should not be used in confined spaces, or in a manner that will allow accumulation of high vapor concentrations. However, for controlled industrial uses when this product is used in confined spaces, heated above ambient temperatures or agitated, the use of explosion proof ventilation equipment is necessary.
Protective Gloves:	Chemical resistant gloves (Neoprene or Natural Rubber).
Eye Protection:	Chemical safety goggles and a face shield.
Other Protective Equipment:	Impervious clothing or boots where needed, eyewash fecility and a safety shower.

SECTION 10 – SPECIAL PRECAUTIONS

Dept. of Labor Storage Category:	Flammable Liquid-Class IB.
Hygienic Practices:	Keep away from heat, sparks and flame. Keep containers closed when not in use. Avoid eye contact. Avoid prolonged or repeated contact with skin. Wash skin with soap and water after contact.
Additional Precautions:	Ground containers when transferring liquid to prevent static accumulation and discharge. Additional information regarding safe handling of products with static accumulation potential can be ordered by contacting the American Petroleum Institute (API) for API Recommended Practice 2003, entitled "Protection Against Ignitions Arising Out of Static, Lighting, and Stray Currents" (American Petroleum Institute, 1720 L Street Northwest, Washington,DC 20005), or the National Fire Protection Association (NFPA) for NFPA 77 entitled "Static Electricity" (National Fire Protection Association, 1 Batterymarch Park, P.O. Box 9101, Quincy, MA 02269-9101).
Empty Container Warning:	"Empty" containers retain residue (liquid and/or vapor) and can be dangerous. Do not pressurize, cut, weld, braze, solder, drill, grind or expose such containers to heat, flame, sparks or other sources of ignition. They may explode and cause injury or death. Do not attempt to clean since residue is difficult to remove. "Empty" drums should be completely drained, properly bunged and promptly returned to supplier or disposed of in an environmentally safe manner and in accordance with governmental regulations.

Figure 3.11 MSDS for acetone (Part II)

3.7 SUMMARY

A written safety and health program helps set out standard safety policies and procedures so that they can be understood and communicated consistently among all stakeholders. Safety and health policy is the first and most important element of a safety and health program because it is a statement of management's commitment to a safe and healthful work environment. Other elements that reflect a company's commitment and safety culture include measurable safety and health goals and clearly defined roles and responsibilities. Project administration that ensures employee safety and health is supported by enforced disciplinary action and accountability, regular safety inspections, accident investigation procedures, record keeping and reporting guidelines, training, medical assistance, emergency response planning, and employee communication.

Phased project planning examines the logical sequence of required construction activities and highlights the potential hazards behind these activities for JHA. A JHA breaks down a construction activity into tasks or steps, lists all potential hazards associated with each task, and suggests proper ways to eliminate or reduce the hazards to an acceptable risk level. During JHA development, when multiple safety approaches are applicable, engineering controls are preferred over other means like administrative controls, PPE, or safe practices. The technical sections of a safety and health program contain safety rules and programs that apply to the project under planning. For this reason, for general contractors, masonry construction and steel erection are typically subcontracted out and are rarely included in safety and health programs. Conversely, "fall protection," "scaffold safety," and "ladder safety" can be expected in almost every safety and health program in the construction industry.

3.8 REVIEW QUESTIONS

1. What strategies can be used to make a safety and health program specific to the project and site conditions?
2. What are the fundamental record keeping and reporting requirements as per OSHA standards?
3. What are the responsibilities of company leaders with respect to creating an effective safety and health program for a construction company?
4. What information should be made available on-site in case of emergencies?
5. Why are job walks an important element in a company safety and health program?
6. Why are engineering controls preferred over other safety means? Are PPEs the last or first resort of safety? Why?
7. What is a JHA, and why is it conducted?
8. What are the project superintendent's responsibilities regarding project site safety planning and enforcement of company safety policies?
9. How often should project site safety inspections be conducted?

10. What type of information should be collected when investigating an accident on a construction project?
11. Why is emergency response planning an essential component of an APP?
12. Why is hazard communication an essential element of an APP?
13. What type of information can be found on a MSDS?

3.9 EXERCISES

1. How can safety and health programs be used to promote construction site safety in practices?
2. Is "Zero Harm" a reasonable safety and health goal for construction projects? Why or why not?
3. What different safety trainings have you received or seen on-site and why are those trainings needed?
4. What are the strategies that can be used to communicate the safety and health program to employees?
5. Select a typical construction activity and apply job hazard analysis to the activity following the example shown in Table 3.1.
6. What incentive programs have you heard of or seen? Were they effective and why?
7. What are the key features of an adequately designed incentive program?

4

OSHA COMPLIANCE

4.1 HISTORY OF THE OSH ACT

Before 1970, comprehensive and uniform provisions that governed worker protection against occupational safety and hazards were nonexistent. The number of job-related accidents was as high as 14,000 deaths and 2 million injuries annually. The productivity lost from disabled workers was 10 times more than that caused by strikes. Although some states had job safety acts, the enforcement was ineffective. There were also issues concerning the varying requirements across different states and the higher operating costs in states that had their own safety acts.

On December 29, 1970, President Richard Nixon signed the OSH Act into law to assure, insofar as possible, safe and healthful working conditions for every working man and woman in the nation and to preserve our nation's human resources. In 1971, based on the OSH Act, Congress created OSHA under the Department of Labor and the National Institute for Occupational Safety and Health (NIOSH) under the Department of Health and Human Services to establish safety standards at the federal level. Because of the OSH Act, the number of job-related accidents fell to about 4000 deaths in the year 2010, even though U.S. employment had almost doubled.

OSHA's assigned mission is to[1]:

- Encourage employers and employees to reduce workplace hazards
- Implement new safety and health programs
- Improve existing safety and health programs
- Encourage research that leads to innovative ways of dealing with workplace safety and health problems
- Establish the rights of employers regarding the improvement of workplace safety and health
- Monitor job-related illnesses and injuries through a system of reporting and record keeping

[1] *All About OSHA*, U.S. Department of Labor, 2010.

- Establish training programs to increase the number of safety and health professionals and to improve their competence continually
- Establish mandatory workplace safety and health standards and enforce those standards
- Provide for the development and approval of state-level workplace safety and health programs
- Monitor, analyze, and evaluate state-level safety and health programs

4.2 OSHA STANDARDS AND THE GENERAL DUTY CLAUSE

The OSHA standards were initially taken from three main sources: consensus standards, proprietary standards, and laws already in effect. Specifically, OSHA has incorporated the consensus standards from the American National Standards Institute (ANSI) and the National Fire Protection Association (NFPA). As a result, a significant portion of the original OSHA standard verbiage read like the "specification language" (e.g., "should") and was modified into the "process regulation language" (e.g., "shall") throughout the years.

The OSHA standards are compiled and published in the Code of Federal Regulations, Title 29 (also referred to as "29 CFR"). Figure 4.1 shows an example OSHA standard on tool safety for the construction industry. Although vertical standards like 29 CFR 1926 designate the OSHA standards for construction particularly, horizontal standards such as 29 CFR 1910 for the general industry also apply because the standards, as the name suggests, are not industry specific and thus have broad coverage. All OSHA standards are freely available online at http://www.osha.gov and its "regulations" page provides a quick access to all the standards. Using the search function on OSHA's main page is another option to locate needed standards but might take multiple tries unless one is sophisticated in forming effective search queries or remembers the specific standard coding numbers.

Part 1926 of the OSHA regulations are entitled "Occupational Safety and Health Standards for Construction." The major subparts of the 1926 standards are:

- Subpart A: General
- Subpart B: General Interpretations
- Subpart C: General Safety and Health Provisions
- Subpart D: Occupational Health and Environmental Control
- Subpart E: Personal Protective and Life Saving Equipment
- Subpart F: Fire Protection and Prevention
- Subpart G: Sign, Signals, and Barricades
- Subpart H: Materials Handling, Storage, Use and Disposal
- Subpart I: Tools—Hand and Power
- Subpart J: Welding and Cutting
- Subpart K: Electrical

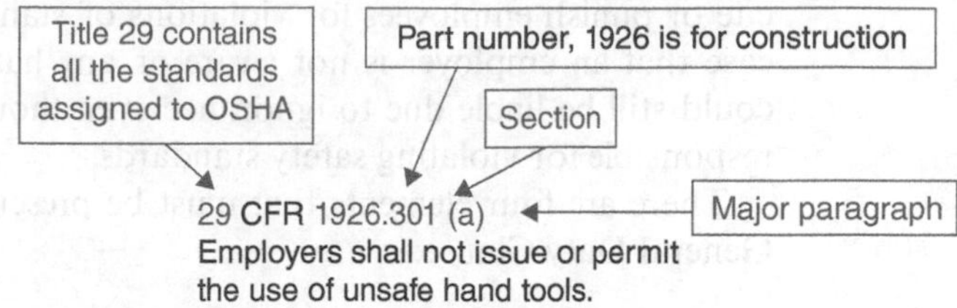

Figure 4.1 Number coding of the OSHA standards

- Subpart L: Scaffolds
- Subpart M: Fall Protection
- Subpart N: Cranes, Derricks, Hoists, Elevators and Conveyors
- Subpart O: Motor Vehicles, Mechanized Equipment, Marine
- Subpart P: Excavations
- Subpart Q: Concrete and Masonry Construction
- Subpart R: Steel Erection
- Subpart S: Underground Construction, Caissons, Cofferdams and Compressed Air
- Subpart T: Demolition Requirements
- Subpart U: Blasting and the Use of Explosives
- Subpart V: Power Transmission and Distributions
- Subpart W: Rollover Protective Structures, Overhead Protection
- Subpart X: Stairways and Ladders
- Subpart Y: Commercial Diving Operations
- Subpart Z: Toxic and Hazardous Substances
- Subpart CC: Cranes and Derricks in Construction

Extracts of the standards that relate to the Fire Station 39 project are shown in Appendix E.

The OSHA standards oversee many areas of concern. Even when there is no specific requirement for a given hazardous situation, the General Duty Clause in the OSH Act (5)(a)(1) should still be followed. The clause states: "Each employer shall furnish to each of his employees employment and a place of employment which are free from recognized hazards that are causing or are likely to cause death or serious physical harm to his or her employees." The General Duty Clause is often cited by the compliance officers when no specific OSHA standard applies to an observed hazard. However, this does not mean that an OSHA compliance officer can use the General Duty Clause to cite any condition that he or she believes is unsafe.

Overall, OSHA places more responsibility upon employers than on employees. While employers are responsible for providing a workplace that is free from hazards that are likely to cause death or injury, they are not responsible for guaranteeing that employees will always follow OSHA requirements. However, employers must do what can be reasonably expected to ensure that the employees work under conditions that meet the OSHA requirements. Additionally, although employees must follow and obey safety rules and regulations, OSHA does not cite or punish employees for violations of standards. On the other hand, in the case that an employer is not aware of any hazardous condition, the employer could still be liable due to ignorance even though it was an employee who was responsible for violating safety standards.

There are four elements that must be present to prove a violation under the General Duty Clause:

- There must be a hazard to which employees are exposed.
- The hazard must be recognized.
- The hazard was causing or was likely to cause death or serious injury.
- The hazard can be corrected in a feasible manner.

4.3 OSHA JURISDICTION AND STATE PROGRAMS

Certain types of organizations or agencies can be exempt from the OSHA requirements because they are considered self-employed or are covered by other federal statues or standards. Examples of such organizations are LLC/partnership corporations (self-employed) and coal mining companies (already covered by the Federal Mine Safety and Health Act). Except for these organizations, other business establishments, even if they are as small as having only one employee, are still covered by the OSHA standards. However, employers of 10 or fewer employees could be exempt from the OSHA record keeping requirement.

A state that has legislative authority and adequate standards development and enforcement capability is encouraged to establish and operate its own safety and health program. However, the standards, enforcement, and compliance requirements in a state program need to be at least as effective as the federal OSHA program. Once a state establishes its own program, OSHA will fund up to 50% of the cost of operating the program.

Taking Washington State as an example, with OSHA's authorization to develop the state's own safety and health program, the Washington legislature enacted the Washington Industrial Safety and Health Act (WISHA). The law authorized the WA L&I, the state's third largest agency, to develop the state's safety and health rules for approval by the state legislature. Once approved, the rules became a part of the Revised Code of Washington (RCW). In some cases, the state rules are even more stringent than the OSHA rules.

By 2012, 27 states and territories—Alaska, Arizona, California, Connecticut, Hawaii, Illinois, Indiana, Iowa, Kentucky, Maryland, Michigan, Minnesota, Nevada, New Jersey, New Mexico, New York, North Carolina, Oregon, Puerto Rico, South Carolina, Tennessee, Utah, Vermont, Virgin Islands, Virginia, Washington, and Wyoming—operated local occupational safety and health plans. These states are shown in Figure 4.2. However, the plans from Connecticut, Illinois, New Jersey, New York, and the Virgin Islands are not complete and cover only public-sector employees at the state and local government levels.

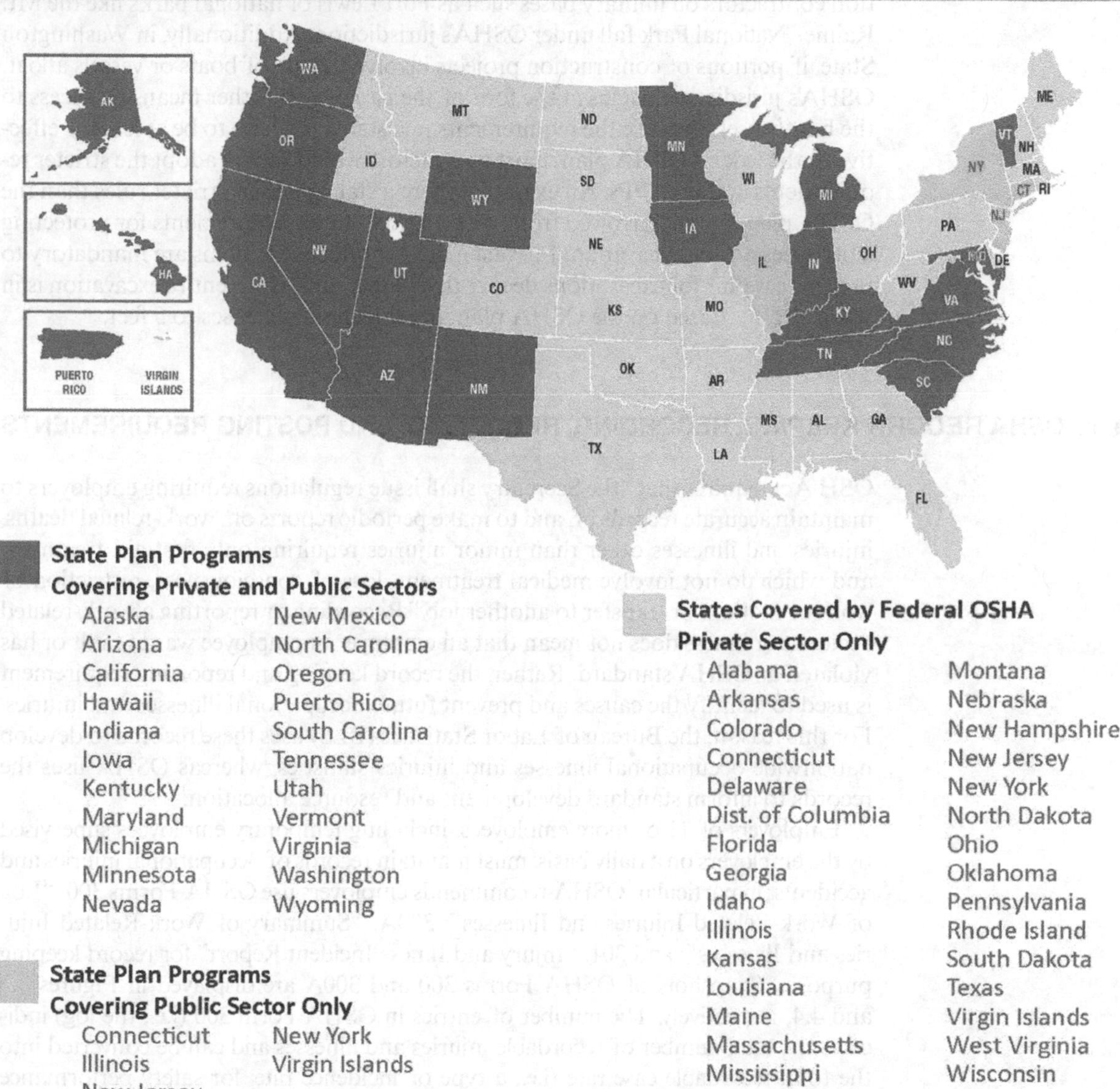

Figure 4.2 States and territories with own occupational safety and health programs

(*Source*: http://www.osha.gov/dcsp/osp/oshspa/oshspa_2009_report.html.)

When a state has its own occupational safety and health plan, OSHA's jurisdiction within the state can be difficult to determine. In Washington State, construction contractors on military bases such as Fort Lewis or national parks like the Mt. Rainier National Park fall under OSHA's jurisdiction. Additionally, in Washington State, if portions of construction projects involve the use of boats or vessels afloat, OSHA's jurisdiction begins at the foot of the gangway or other means of access to the boat/vessel. Because the requirements in a state plan have to be at least as effective as the federal OSHA plan, most contractors would simply adopt the stricter requirements in their APPs. An example where a state plan has stricter rules than the OSHA plan can be borrowed from Washington State's requirements for protecting employees during excavation. In Washington, protective systems are mandatory to prevent cave-ins for excavations deeper than 4 feet (unless the entire excavation is in stable rocks). Based on the OSHA plan, the threshold increases to 5 feet.

4.4 OSHA RECORD KEEPING, RECORDING, REPORTING, AND POSTING REQUIREMENTS

OSH Act requires that "the Secretary shall issue regulations requiring employers to maintain accurate records of, and to make periodic reports on, work-related deaths, injuries and illnesses other than minor injuries requiring only first aid treatment and which do not involve medical treatment, loss of consciousness, restriction of work or motion, or transfer to another job." Recording or reporting a work-related accident or illness does not mean that an employer or employee was at fault or has violated an OSHA standard. Rather, the record keeping and reporting requirement is used to identify the causes and prevent future occupational illnesses and injuries. For this reason, the Bureau of Labor Statistics (BLS) uses these records to develop nationwide occupational illnesses and injuries statistics, whereas OSHA uses the records to inform standard development and resource allocation.

Employers of 11 or more employees, including temporary employees supervised by the employers on a daily basis, must maintain records of occupational injuries and accidents. In particular, OSHA recommends employers use OSHA Forms 300, "Log of Work-Related Injuries and Illnesses"; 300A, "Summary of Work-Related Injuries and Illnesses"; and 301, "Injury and Illness Incident Report" for record keeping purposes. Snapshots of OSHA Forms 300 and 300A are displayed in Figures 4.3 and 4.4, respectively. The number of entries in OSHA Form 300 (i.e., the log) indicates the total number of recordable injuries and illnesses and can be converted into the total recordable case rate (i.e., a type of incidence rate, for safety performance comparison). Incidence rate is a statistic for measuring safety performance and for comparing the safety performance of different groups of employers. In business terms, it is commonly used for job prequalification, field safety performance monitoring and benchmarking, and determination of workers' compensation premiums. The following formula indicates how to calculate the total recordable case rate:

$$\text{Total recordable case rate} = \frac{\text{total injuries and illnesses} \times 200{,}000}{\text{number of hours worked by all employees}}$$

OSHA's Form 300 (Rev. 01/2004)

Log of Work-Related Injuries and Illnesses

Attention: This form contains information relating to employee health and must be used in a manner that protects the confidentiality of employees to the extent possible while the information is being used for occupational safety and health purposes.

Year 20___

U.S. Department of Labor
Occupational Safety and Health Administration

Form approved OMB no. 1218-0176

You must record information about every work-related death and about every work-related injury or illness that involves loss of consciousness, restricted work activity or job transfer, days away from work, or medical treatment beyond first aid. You must also record significant work-related injuries and illnesses that are diagnosed by a physician or licensed health care professional. You must also record work-related injuries and illnesses that meet any of the specific recording criteria listed in 29 CFR Part 1904.8 through 1904.12. Feel free to use two lines for a single case if you need to. You must complete an Injury and Illness Incident Report (OSHA Form 301) or equivalent form for each injury or illness recorded on this form. If you're not sure whether a case is recordable, call your local OSHA office for help.

Establishment name ______________

City ______________ State ______________

Identify the person			Describe the case			Classify the case: CHECK ONLY ONE box for each case based on the most serious outcome for that case:				Enter the number of days the injured or ill worker was:		Check the "Injury" column or choose one type of illness:					
								Remained at Work				(M)					
(A) Case no.	(B) Employee's name	(C) Job title (e.g., Welder)	(D) Date of injury or onset of illness	(E) Where the event occurred (e.g., Loading dock north end)	(F) Describe injury or illness, parts of body affected, and object/substance that directly injured or made person ill (e.g., Second degree burns on right forearm from acetylene torch)	Death (G)	Days away from work (H)	Job transfer or restriction (I)	Other record-able cases (J)	Away from work (K)	On job transfer or restriction (L)	Injury (1)	Skin disorder (2)	Respiratory condition (3)	Poisoning (4)	Hearing loss (5)	All other illnesses (6)
			/ month/day			☐	☐	☐	☐	___ days	___ days	☐	☐	☐	☐	☐	☐
			/ month/day			☐	☐	☐	☐	___ days	___ days	☐	☐	☐	☐	☐	☐
			/ month/day			☐	☐	☐	☐	___ days	___ days	☐	☐	☐	☐	☐	☐
			/ month/day			☐	☐	☐	☐	___ days	___ days	☐	☐	☐	☐	☐	☐
			/ month/day			☐	☐	☐	☐	___ days	___ days	☐	☐	☐	☐	☐	☐
			/ month/day			☐	☐	☐	☐	___ days	___ days	☐	☐	☐	☐	☐	☐
			/ month/day			☐	☐	☐	☐	___ days	___ days	☐	☐	☐	☐	☐	☐
			/ month/day			☐	☐	☐	☐	___ days	___ days	☐	☐	☐	☐	☐	☐
			/ month/day			☐	☐	☐	☐	___ days	___ days	☐	☐	☐	☐	☐	☐
			/ month/day			☐	☐	☐	☐	___ days	___ days	☐	☐	☐	☐	☐	☐
			/ month/day			☐	☐	☐	☐	___ days	___ days	☐	☐	☐	☐	☐	☐
			/ month/day			☐	☐	☐	☐	___ days	___ days	☐	☐	☐	☐	☐	☐
			/ month/day			☐	☐	☐	☐	___ days	___ days	☐	☐	☐	☐	☐	☐
					Page totals➤	___	___	___	___	___	___	___	___	___	___	___	___
												Injury (1)	Skin disorder (2)	Respiratory condition (3)	Poisoning (4)	Hearing loss (5)	All other illnesses (6)

Be sure to transfer these totals to the Summary page (Form 300A) before you post it.

Page ___ of ___

Public reporting burden for this collection of information is estimated to average 14 minutes per response, including time to review the instructions, search and gather the data needed, and complete and review the collection of information. Persons are not required to respond to the collection of information unless it displays a currently valid OMB control number. If you have any comments about these estimates or any other aspects of this data collection, contact: US Department of Labor, OSHA Office of Statistical Analysis, Room N-3644, 200 Constitution Avenue, NW, Washington, DC 20210. Do not send the completed forms to this office.

Figure 4.3 OSHA's Form 300, log of work-related injuries and illnesses

OSHA's Form 300A (Rev. 01/2004)

Summary of Work-Related Injuries and Illnesses

Year 20____

U.S. Department of Labor
Occupational Safety and Health Administration

Form approved OMB no. 1218-0176

All establishments covered by Part 1904 must complete this Summary page, even if no work-related injuries or illnesses occurred during the year. Remember to review the Log to verify that the entries are complete and accurate before completing this summary.

Using the Log, count the individual entries you made for each category. Then write the totals below, making sure you've added the entries from every page of the Log. If you had no cases, write "0."

Employees, former employees, and their representatives have the right to review the OSHA Form 300 in its entirety. They also have limited access to the OSHA Form 301 or its equivalent. See 29 CFR Part 1904.35, in OSHA's recordkeeping rule, for further details on the access provisions for these forms.

Number of Cases

Total number of deaths	Total number of cases with days away from work	Total number of cases with job transfer or restriction	Total number of other recordable cases
______	______	______	______
(G)	(H)	(I)	(J)

Number of Days

Total number of days away from work	Total number of days of job transfer or restriction
______	______
(K)	(L)

Injury and Illness Types

Total number of . . .
(M)

(1) Injuries	______	(4) Poisonings	______
		(5) Hearing loss	______
(2) Skin disorders	______	(6) All other illnesses	______
(3) Respiratory conditions	______		

Post this Summary page from February 1 to April 30 of the year following the year covered by the form.

Public reporting burden for this collection of information is estimated to average 58 minutes per response, including time to review the instructions, search and gather the data needed, and complete and review the collection of information. Persons are not required to respond to the collection of information unless it displays a currently valid OMB control number. If you have any comments about these estimates or any other aspects of this data collection, contact: US Department of Labor, OSHA Office of Statistical Analysis, Room N-3644, 200 Constitution Avenue, NW, Washington, DC 20210. Do not send the completed forms to this office.

Establishment information

Your establishment name ______________________

Street ______________________

City ______________ State ____ ZIP ______

Industry description (*e.g., Manufacture of motor truck trailers*) ______________________

Standard Industrial Classification (SIC), if known (*e.g., 3715*) __ __ __ __

OR

North American Industrial Classification (NAICS), if known (e.g., 336212) __ __ __ __ __ __

Employment information (*If you don't have these figures, see the Worksheet on the back of this page to estimate.*)

Annual average number of employees ______

Total hours worked by all employees last year ______

Sign here

Knowingly falsifying this document may result in a fine.

I certify that I have examined this document and that to the best of my knowledge the entries are true, accurate, and complete.

______________________ Company executive ______ Title

( ) ______________ Phone / / Date

Figure 4.4 OSHA's Form 300A, summary of work-related injuries and illnesses

Because an incidence rate is defined as the number of recordable injuries and illnesses occurring among 100 full-time workers over one year, the 200,000 value in the formula represents the number of hours worked by 100 employees working 40 hours per week, 50 weeks per year, and provides the standard base for calculating incidence rates.

Another type of incidence rate can be calculated based only on the number of cases that involve days away from work and days of restricted work activity or job transfer. The rate is called the DART incidence rate and is calculated using the following formula:

$$\text{DART incidence rate} = \frac{(\text{DA} + \text{JT})(200{,}000)}{\text{number of hours worked by all employees}}$$

where DA is the number of days away from work cases and JT is the number of job transfer or restriction cases.

When an employer classifies a work-related and recordable injury or illness case in OSHA Form 300 (i.e., the log), it is important that the employer only selects the most serious outcome for that case. Therefore, if a case involves both days away from work and job transfer/restriction, the case is classified as a "days away from work" case and only column H has to be selected in OSHA Form 300.

Record keeping forms are maintained on a calendar-year basis and must be kept for five years at the place of business and be available for inspection by federal or state OSHA representatives. Employers of 10 or fewer employees could be exempt from the OSHA recording keeping requirement, although they still need to report fatalities and multiple hospitalization incidents.

As per 29 CFR 1904.7, the general record keeping criterion states that an injury or illness must be recorded if it is work-related and results in one or more of the following situations:

- Death
- Loss of consciousness
- Days away from work
- Restricted work activity
- Job transfer
- Medical treatment beyond first aid
- Diagnosed by a physician or other licensed health care professional
- Cancer-related illnesses
- Chronic irreversible diseases
- Fractured or cracked bone
- Punctured eardrum
- Other special conditions

Situations such as when an employee is contaminated (e.g., through cuts) with another person's blood or other potentially infectious materials or has experienced a standard threshold shift (STS) in hearing might be qualified as "other special conditions" for record keeping purposes.

An injury or illness is considered work related if an event or exposure in the work environment caused or contributed to the condition or significantly aggravated a preexisting condition. The work environment is not limited to job locations only. Injuries or illnesses that occur while employees are traveling can still be considered work related if the employees were engaging in work activities (e.g., traveling from one job site to another) at the time of the injuries/illnesses.

When an injury or illness involves (1) the death of any worker from a work-related incident or (2) in-patient hospitalization of three or more workers as the result of a work-related incident, employers must report to OSHA within 8 hours of learning about the injury or illness. Some state plans have stricter rules on work-related injury/illness reporting. For example, in Washington State, an employer is required to report any workplace accident that results in any fatality or the hospitalization of any employee.

Besides the record keeping and reporting requirements, employers must also post the following materials in workplace locations that are accessible to workers:

- Job safety and health protection workplace poster informing employees of their rights and responsibilities under the OSH Act
- Summaries of petitions for variances from standards or record keeping procedures
- Copies of all OSHA citations for violations of standards (Copies must remain posted at or near the location of alleged violations for three days or until the violations are abated, whichever is longer.)
- OSHA Form 300A, "Summary of Work-Related Injuries and Illnesses" (posted from February 1 to April 30 of the year following the year covered by the form)

If an employee is discriminated against or punished for reporting work-related fatalities, injuries, illnesses, or concerns, the employee can file a complaint with OSHA through the Whistle Blower Protection Program.

4.5 OSHA INSPECTIONS AND CITATIONS

The 1970 OSH Act also authorized OSHA to conduct workplace inspections and investigations to determine if employers are complying with the OSHA standards. Except under special circumstances, inspections mostly take place without prior notice. If an employer refuses to admit an OSHA compliance officer or attempts to interfere with the inspection, the compliance officer could obtain a warrant to inspect the employer based on the OSH Act. Normally, inspections are scheduled based on the following priorities:

- *Imminent Danger.* It is any condition where there is a reasonable belief that a danger exists that can be expected to cause death or serious injury immediately or before the danger can be eliminated through normal enforcement procedures. This may include health hazards such as dangerous fumes or dust.

- *Catastrophes and Fatal Accidents.* Accidents that result in a fatality or hospitalization of three or more workers.
- *Complaints and Referrals.* Complaints from employees or others of alleged safety regulation violations or unsafe working conditions.
- *Programmed High-Hazard Inspections.* Inspections aimed at specific high-hazard workplaces based on past death, injury, or illness incidence rates.
- *Follow-Up Inspections.* Inspections to verify whether previously cited violations have been corrected.

An imminent danger is a hazardous situation in which fatalities or serious injuries are likely to happen immediately before the hazard can be eliminated. An example is a worker in an unprotected trench excavation that is more than 5 feet deep. It thus receives a higher priority for inspection than workplaces that just had catastrophes or fatal accidents. Employees may file a confidential complaint with OSHA to request workplace inspections. Programmed inspections aim at high-hazard industries. For new construction projects, programmed inspections take place after major construction activities commence and when projects are 30 to 60% complete.

The inspection process begins with the compliance officer displaying his or her credentials and making an opening conference. The officer then walks through and inspects the workplace, accompanied by selected employee and employer representatives. In addition to identifying unsafe or unhealthful working conditions, during the walkthrough inspection the officer will also review safety and health records, examine Accident Prevention and Hazard Communication Programs, and determine if the employer satisfies OSHA's record keeping requirements. The inspection process ends with a closing conference to discuss all identified unsafe/unhealthful conditions and violations. The compliance officer will inform the employer of appeal rights but will not discuss any proposed penalty during inspection.

When a compliance officer identifies violations, OSHA might issue citations that state penalties, the violated standards, and deadlines for corrections. Issuing warnings in lieu of citations is not an option. Generally, a violation means that there is noncompliance with the OSHA standards and that one or more employees are exposed to hazards as a result. On multiemployer workplaces, such as most of the projects in construction, citations shall normally be issued only to employers whose employees are exposed to hazards, although the primary controlling employer (such as the general contractor with responsibility for correcting the hazards) may be cited instead. An example citation is provided in Figure 4.5 to show the common information in a citation.

There are five types of violations that may be cited:

- *Other-Than-Serious* (\$0−\$1000). A violation that has a direct relationship to job safety and health, but probably would not cause death or serious physical harm. For example, an employer fails to comply with OSHA's information posting requirement.

U.S. Department of Labor
Occupational Safety and Health Administration

Inspection Number:
Inspection Dates:
Issuance Date:

Citation and Notification of Penalty

Company Name: ABC Company, Inc.
Inspection Site: Anywhere USA

Citation 2 Item 1d Type of Violation: Willful

29 CFR 1910.1000(c): Employees were exposed to crystalline silica (quartz) and respirable dust, listed in Table Z-3, in excess of the 8-hour Time-Weighted Average concentrations listed for those materials:

a) On June 21, 1996, a sandblaster working in the shed was exposed to respirable dust containing 41.6% crystalline silica (quartz) at a Time-Weighted Average level of 34.983 mg/m³, which is approximately 153 times the Permissible Exposure Limit for crystalline silica of 0.229 mg/m³ and approximately 7 times the Permissible Exposure Limit for respirable dust of 5 mg/m³. The exposure level is derived from two samples collected over a 444-minute period; the calculations include a zero value for the 36 minutes not sampled.

b) On June 21, 1996, a sandblaster working in the shed was exposed to respirable dust containing 44.0% crystalline silica (quartz) at a Time-Weighted Average level of 10.073 mg/m³, which is approximately 46 times the Permissible Exposure Limit for crystalline silica of 0.217 mg/m³ and approximately 2 times the Permissible Exposure Limit for respirable dust of 5 mg/m³. The exposure level is derived from two samples collected over a 427-minute period; the calculations include a zero value for the 53 minutes not sampled.

c) On June 21, 1996, a sandblaster working in the booth was exposed to respirable dust containing 50.7% crystalline silica (quartz) at a Time-Weighted Average level of 28.602 mg/m³, which is approximately 151 times the Permissible Exposure Limit for crystalline silica of 0.190 mg/m³ and approximately 5.7 times the Permissible Exposure Limit for respirable dust of 5 mg/m³. The exposure level is derived from two samples collected over a 478-minute period; the calculations include a zero value for the 2 minutes not sampled.

d) On June 21, 1996, a sandblast assistant working in the vicinity of the shed was exposed to respirable dust containing 46.7% crystalline silica (quartz) at a Time-Weighted Average level of 2.042 mg/m³, which is approximately 10 times the Permissible Exposure Limit for crystalline silica of 0.205 mg/m³. The exposure level is derived from two samples collected over a 458-minute period; the calculations include a zero value for the 22 minutes not sampled.

e) On June 21, 1996, a sandblaster working in the shed in the morning and patching asphalt paths in the afternoon was exposed to respirable dust containing 27.0% crystalline silica (quartz) at a Time-Weighted Average level of 1.033 mg/m³, which is approximately 3 times the Permissible Exposure Limit for crystalline silica of 0.345 mg/m³. The exposure level is derived from two samples collected over a 461-minute period; the calculations include a zero value for the 19 minutes not sampled.

Date By Which Violation Must be Abated: --/--/--

See pages 1 through 3 of this Citation and Notification of Penalty for information on employer and employee rights and responsibilities.

Citation and Notification of Penalty OSHA 2 (Rev. 6/93)

Figure 4.5 OSHA citation example

- *Serious* ($1500–$7000). A violation where there is a substantial probability that death or serious physical harm could result. For example, employees walking/working on surfaces more than 6 feet above lower levels are not protected from falling through holes (including skylights).
- *Willful* ($5000–$70,000). A violation that the employer intentionally and knowingly commits.
- *Repeat* (*Up to* $70,000). A violation of any standard, regulation, rule, or order where, upon reinspection, a substantially similar violation is found and the original citation has become a final order.
- *Failure To Abate* ($7000 *per day*). Failure to correct a prior violation may bring a civil penalty of up to $7000 for each day.

For other-than-serious and serious violations, OSHA may adjust a penalty based on the employer's demonstrated good faith, previous inspection history, size of business, and gravity of violation. As mentioned previously, the employer must post copies of the citations at or near the location of alleged violations for three days, or until the violations are abated, whichever is longer.

A citation is only the compliance officer's belief that a violation exists. An employer may officially appeal the citation in writing (i.e., notice of contest) within 15 days of the receipt of the citation in order to contest the citation, the abatement period, or the penalty. Employees may not contest a citation but may contest the time allotted to correct a violation (e.g., if it is too long) or employer's "Petition for Modification of Abatement." If an inspection was initiated because of an employee's complaint, the employee may request a review on why a citation was not issued.

4.6 OSHA SERVICES AND PROGRAMS

Free consultation services are available from OSHA for employers who need help to establish and maintain a safe and healthful workplace. The consultation services help an employer survey its work site hazards and evaluate existing safety and health management systems without any penalties or citations attached. The employer's only obligation is to correct all identified serious hazards within the agreed-upon time period. In the case where no such system is available, the services can assist the employer to develop one.

Employers with outstanding safety achievements can be recognized by OSHA's Voluntary Protection Program (VPP) at three levels: Star, Merit, and Demonstration. For the employers who are more interested in cooperative relationships with groups of employers and employees, such as trade unions and professional associations, they can participate in OSHA's Strategic Partnership Program.

Through its Susan Harwood Grant, OSHA also funds initiatives that aim to conduct safety and health training. An organization could apply for the grant to develop safety and health training materials, to deliver training sessions, or to increase its training capacity by strengthening the number and knowledge of in-house staff.

4.7 SUMMARY

The OSHA standards are compiled and published in the Code of Federal Regulations, Title 29, and cover many safety and health areas of concern. When there is no specific requirement that applies for a given hazardous situation, the General Duty Clause should still be followed. Certain types of organizations or agencies can be exempt from the OSHA requirements because they are considered self-employed or are covered by other federal statutes or standards. Other business establishments are still covered by the OSHA standards even if they are as small as having only one employee. States can establish and operate their own job safety and health programs as long as the programs are at least as effective as the federal OSHA program.

Employers of 11 or more employees must maintain records of occupational injuries and accidents. Typically, this can be done by using OSHA Forms 300 ("Log"), 300A ("Summary"), and 301 ("Incident Report"). Record keeping forms are maintained on a calendar-year basis and must be kept for five years at the place of business and be available for inspection by federal or state OSHA representatives. OSHA's general record keeping criterion states that an injury or illness must be recorded if it is work related and results in deaths or injuries and illnesses other than minor injuries requiring only first-aid treatment and which do not involve medical treatment, loss of consciousness, restriction of work or motion, or transfer to another job. When an injury or illness involves (1) death of any worker from work-related incident or (2) in-patient hospitalization of three or more workers as result of work-related incident, employers must report to OSHA within 8 hours of learning about the injury or illness.

Besides the record keeping and reporting requirements, an employer must also post information on worker's OSHA rights and responsibilities as well as the employer's safety and health performance at workplace locations that can be accessible to workers. OSHA compliance officers conduct workplace inspections and investigations to determine if employers are complying with the OSHA standards. Normally, inspections are scheduled using the following priorities: "imminent danger," "catastrophes and fatal accidents," "complaints and referrals," "programmed inspections," and "follow-up inspections." When a compliance officer finds violations, OSHA might issue citations that state penalties, the violated standards, and deadlines for corrections. The five types of violations that may be cited are "other than serious," "serious," "willful," "repeat," and "failure to abate." A citation is only the compliance officer's belief that a violation exists and an employer may officially appeal the citation in writing through a notice of contest within 15 days of receipt of the citation. OSHA also maintains consultation services and special programs such as the VPP and Strategic Partnership programs to promote workplace safety and health on top of its Susan Harwood Training Grant.

4.8 REVIEW QUESTIONS

1. Are business establishments with 10 or fewer employees exempt from the OSHA requirements?
2. What are some of the responsibilities employers have related to OSHA record keeping?
3. What are OSHA's record keeping criteria?
4. What are OSHA's reporting requirements?

5. What are OSHA's information posting requirements that keep employees informed?
6. What is the general process of an OSHA workplace inspection?
7. What is the difference between OSHA's vertical and horizontal standards?
8. How are OSHA work site inspections prioritized?

4.9 EXERCISES

1. Discuss if the following injuries are recordable:
   (a) An employee sprained his wrist at a job site and was given Advil before returning to his task.
   (b) An employee sprained his wrist at a job site and received medication as well as light-duty tasks.
   (c) The same employee, while still being on medication for the sprained wrist, complained about the wrist pain after moving some light materials at the job site.
   (d) The same employee, after being fully recovered, complained about the wrist pain after moving some materials at the job site.
2. Discuss what hazardous scenarios in construction can be considered as imminent danger.
3. Practice recording the following workplace accident in an OSHA Form 300. The accident involves James Pat, an electrician who fell off a ladder and hurt his shoulder on October 14, 2012, while working in the 25th floor cafeteria. James had to be away from work for seven days and was on lighter duty for another two weeks before he fully recovered.
4. Go to OSHA's website and locate its record keeping standards to verify the criteria for recording occupational-related hearing loss injuries on the OSHA Form 300 log.

5

SAFETY FOR PROJECT START-UP

5.1 PROJECT OVERVIEW

Fire Station 39 (FS39), located in Lake City, Washington, will be used as a case example throughout the book to discuss, in phases, occupational hazards that are typical to a construction project. The example project is an 11,200-ft^2, structural steel frame building owned by the City of Seattle. The new station building was intended to replace the old 3200-ft^2 fire station and includes a two-story station house, a one-story apparatus bay, and a one-story storage wing. The station also consists of an enclosed yard and a second-floor terrace.

The project was valued at approximately $3.5 million and its construction took 11 months (April 2009 to March 2010) starting from Notice to Proceed. As required by the owner, the project was certified through the U. S. Green Building Council's Leadership in Energy and Environmental Design (LEED) Green Building Rating System Certification Program. The project went above and beyond the City of Seattle's LEED Silver requirement and achieved a LEED Gold certification at its completion. In particular, the project features a rainwater harvesting cistern, a stormwater garden, and a galvanized metal sculpture that receives the stormwater runoff and transports it to the cistern. By design, when the cistern fills up, excess rainwater spills into the stormwater garden at the south end. Figure 5.1 illustrates the 28-foot-tall, freestanding metal sculpture in the rain garden of the fire station.

The project site was located in the upmost northeast corner of Seattle in Lake City, roughly 9 miles from downtown Seattle, and was easily accessible by the main road in the area, Lake City Way. Although the site was within the boundary planned for the area's higher density residential/commercial district and was close to the Lake City's civic buildings, it was on a relatively loose lot considering its surroundings. Specifically, the site was at the northeast corner of its nearest street intersection (i.e., 28th Ave. NE and NE 127 St.), adjoining the old fire station on the east and a six-story apartment building on the north. In addition, two auto repair/dealer shops sat across the roads from the south and west ends of the site. There was no heavy foot or vehicular traffic near the site and street parking, although not ample, was possible west of 28th Ave. NE on NE 127 St. Visible overhead power lines ran along the roads on the opposite sides away from the site. Figure 5.2 displays the bird's-eye view of the site location.

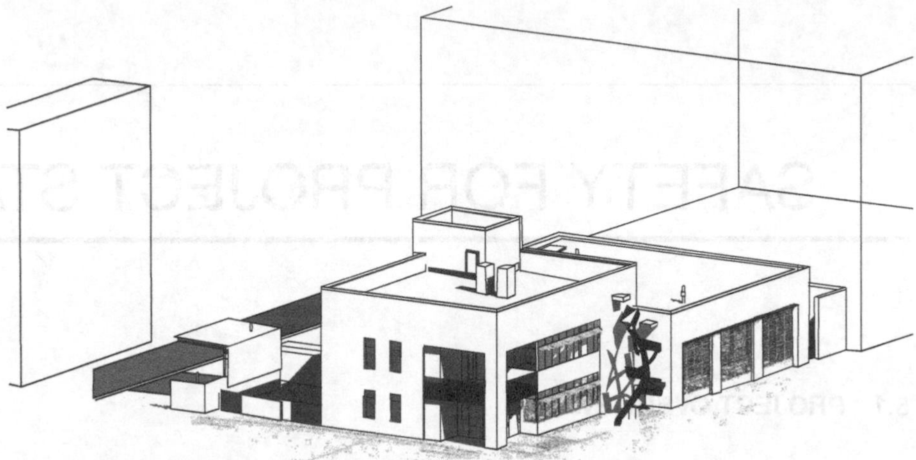

Figure 5.1 South view of FS39 featuring the 28-foot metal sculpture
(Courtesy of the Miller Hull Partnership, LLP.)

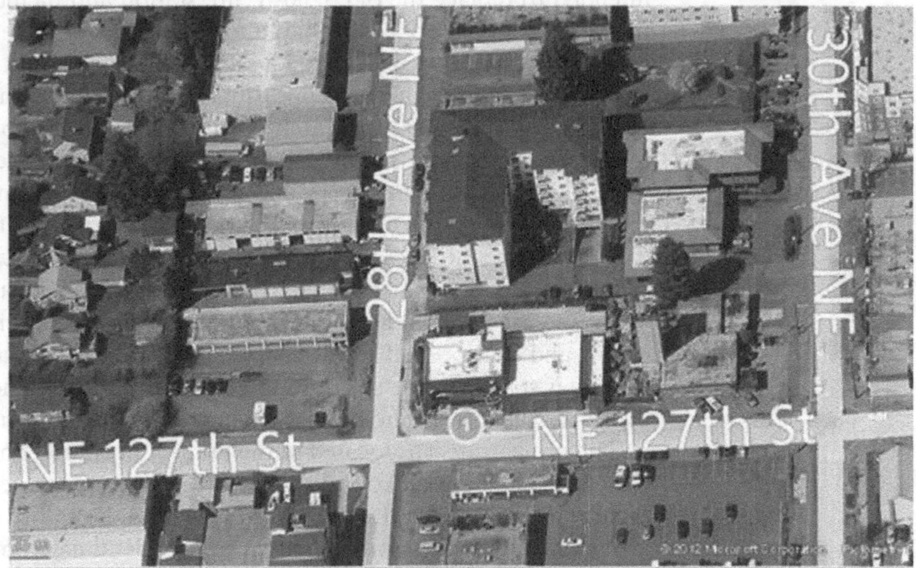

Figure 5.2 Bird's-eye view of the FS39 site location

5.2 PREPROJECT PLANNING

The purpose of preproject planning is to anticipate and address any potential problems or logistical requirements before actual construction takes place. It is an extremely experience-driven process and benefits tremendously from the planner's deep understanding of construction means and methods. As a powerful tool that facilitates communication among all project stakeholders, building information modeling (BIM) especially helps visualize the construction sequencing and site challenges. The City of Seattle specifically requested that the project design firm create a building information model for FS39 and to share the model with the general contractor. For copyright reasons, a separate, simplified building information model has been created for this book to show the major construction activities needed for FS39, and a copy of the overall project schedule is provided in Appendix C as a reference. Figures 5.3, 5.4, 5.5, 5.6, and 5.7 display snapshots

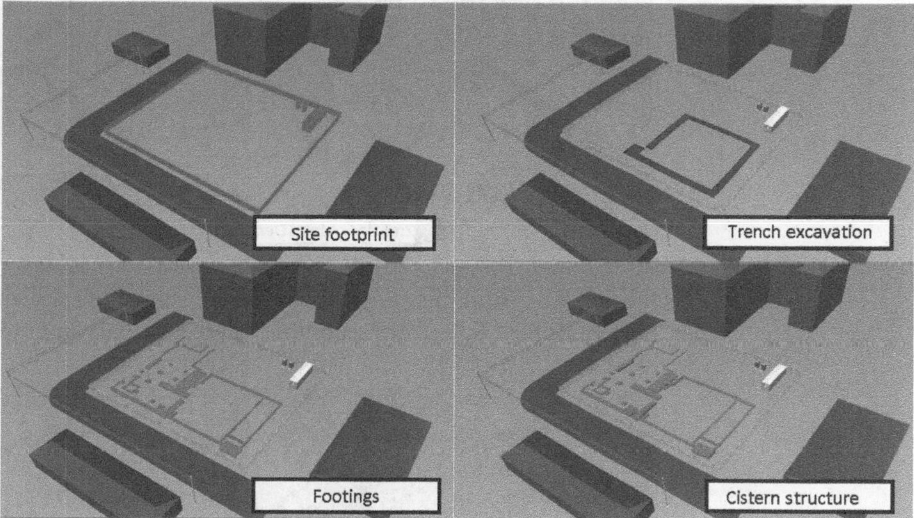

Figure 5.3 Building information model snapshots showing main project activities for FS39 (from site fencing to cistern)

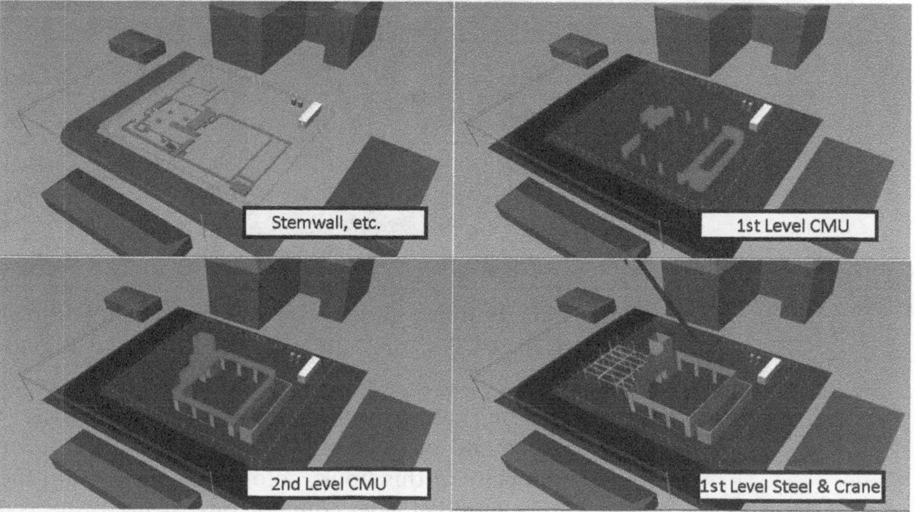

Figure 5.4 Building information model snapshots showing main project activities for FS39 (from stem wall to first-level steel)

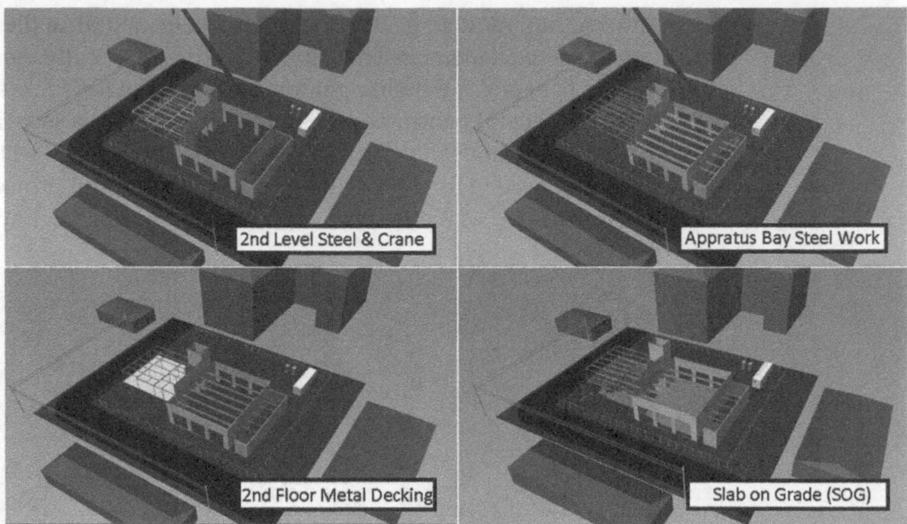

Figure 5.5 Building information model snapshots showing main project activities for FS39 (from second-level steel to slab on grade)

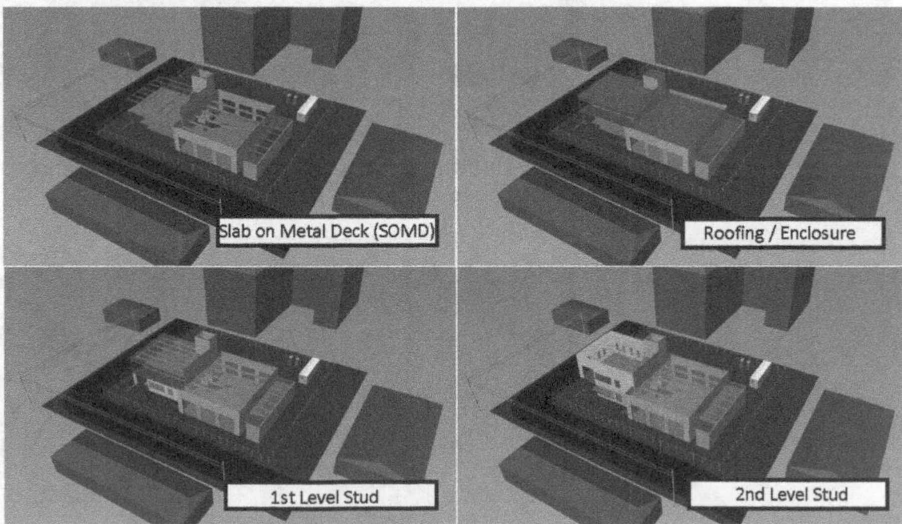

Figure 5.6 Building information model snapshots showing main project activities for FS39 (from slab on metal decking to second-level stud)

of the building information model for these activities. While the construction photographs and pictures of the building information model are presented in black and white in this book, color versions of both the photographs and pictures can be found at wiley.com/go/constructionprojectsafety.

As the building information model illustrates, the deepest excavation of the site was near the underground cistern structure and there were no column footings

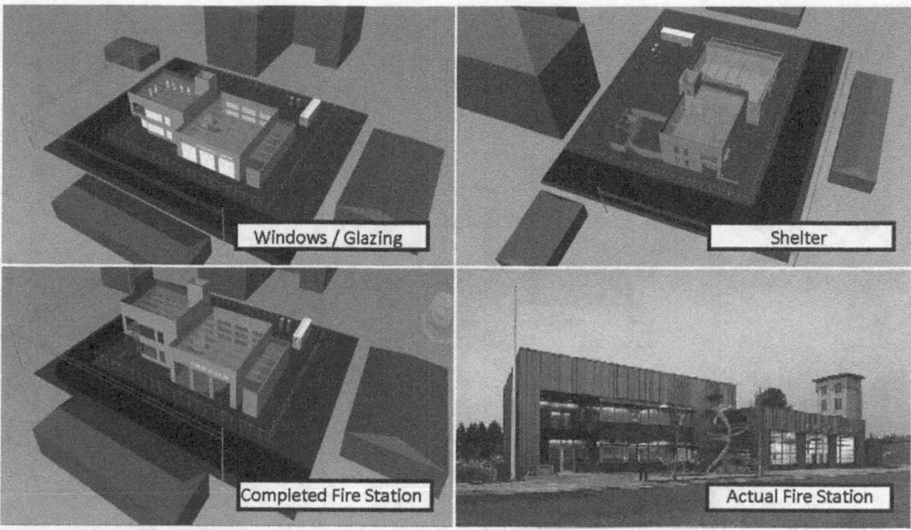

Figure 5.7 Building information model snapshots showing main project activities for FS39 (from glazing to the completed building) and an actual image of the completed building

(Courtesy of the Miller Hull Partnership, LLP.)

under the apparatus bay building, except the stem walls that went around the perimeter of the building. Therefore, logistically, it would make the most sense to start the excavation and foundation work at the cistern and apparatus bay stem walls and then gradually move to the remaining stem walls and column footings under the station house. One benefit of scheduling the sequence of excavation/ foundation work as such is that, once the stem walls under the apparatus bay are completed, the masonry walls can be put up for the steelwork.

The building information model also indicates that the north and east strips of the site did not have major structures above ground (except some masonry work for a small fueling station below the northwest corner) and could accommodate the needs for material delivery/storage/hoisting and job trailer placement. For this reason, the northwest and southeast corners of the site were considered to be the best locations for site entry and egress. Figure 5.8 shows the site logistic plan and how the site can be efficiently utilized. Decisions made when developing a site logistic plan will provide a framework for all the operations on-site and influence every aspect of project execution.

The site trailer and tool shack were situated in the northeast corner so that they did not have to be relocated until the project was close to its completion. Because reconfiguring the old fire station parking lot was a part of the project scope, the site fencing extended to the edge of the parking lot with a gate on the southeast corner for the general contractor's company vehicles to enter the parking area. A mobile crane for hosting structural steel and heavy mechanical systems could be set up north of the fire station structure, keeping a safe clearance between the

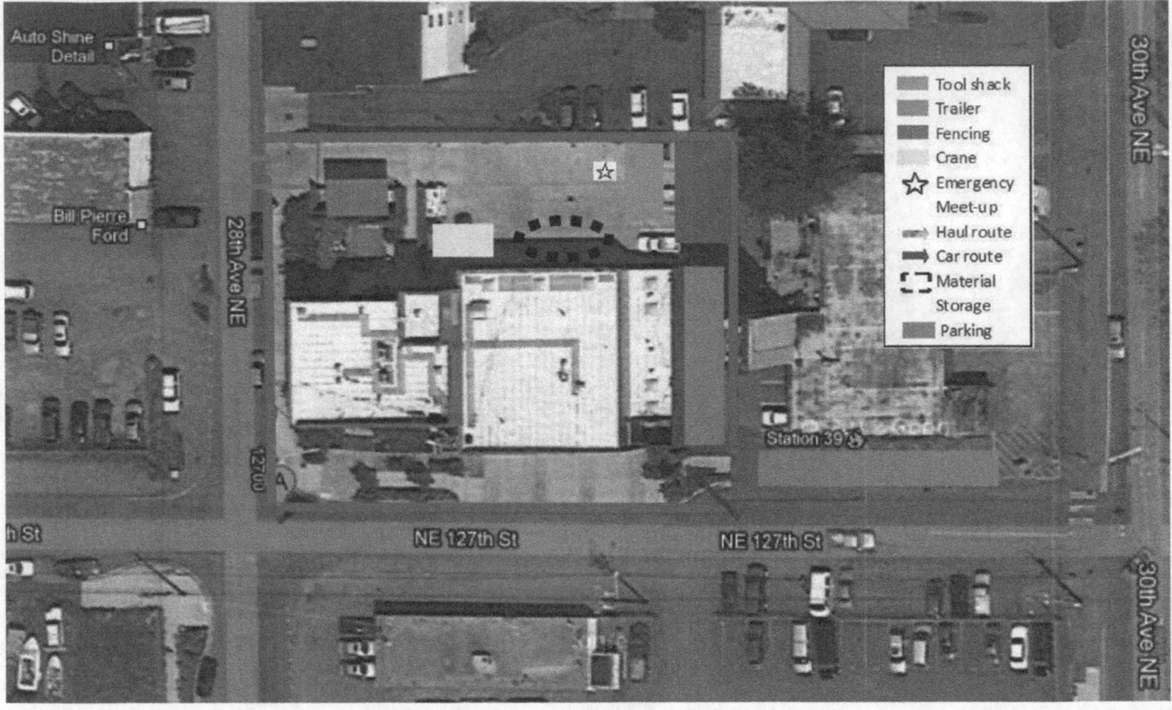

Figure 5.8 FS39 site logistic plan
(Courtesy of the Miller Hull Partnership, LLP.)

crane and the overhead power lines that ran along 28th NE Ave. and NE 127th St. The emergency meet-up point was close to the tool shack where the emergency contact information and first-aid took kit were located. The material hauling and delivery would enter the site from the northwest corner and materials which were not subject to weather conditions could be temporarily stored north of the fire station structure. Finally, two designated parking areas, one inside the site fencing for the general contractor's company vehicles as well as visitors and one just south of the old fire station, should accommodate the day-to-day parking need for the project. Signage for informing pedestrians and neighboring residents should be placed around the site fencing and signage indicating the requirement to wear PPE before entering the site should be placed near the two site entrances.

5.3 PHASED PROJECT-SPECIFIC ACCIDENT PREVENTION PLAN

Preproject planning helps define the construction sequencing and site logistics. After major construction activities are identified, they can be grouped into major phases for the analysis of project-specific hazards to include the use of any hazardous materials. The outcomes of this analysis serve as pointers to highlight particular safe practices, rules, and precautions that should be covered by the technical sections in a project-specific APP. For the FS39 project, activities illustrated from Figures 5.3 to 5.7 can be grouped into four phases: mobilization, earthwork and foundation, superstructure, and exterior enclosure. Including the interior enclosure phase, which is not shown in the building information model, there are five phases to be analyzed. The following discussion walks through each of the five phases and determines each phase's project-specific hazards from a general contractor's point of view.

Ensuring the safety of the general public is always a topic worthy of conversation, even before a project commences. Although FS39 was located in an urban setting, the project did not interfere with heavy pedestrian or vehicular traffic. For this reason, public safety did not seem to pose the biggest concern for this project but regular safety measures such as site fencing, signage, and flaggers (as needed) were still necessary. Before site clearing, underground utilities should be relocated by the responsible party as per contract specifications. The need to keep a minimum clearance between the nearby overhead power lines and site personnel as well as equipment should be clearly outlined in the APP, with corresponding technical sections laying out specific clearance requirements. Temporary erosion and sedimentation control is often necessary, as per project specifications, as it helps maintain the environmental and health aspects of the project.

Prior to any earthwork or excavation, the project team should request the utility locating services to identify existing underground utilities. Some demolition to the existing site pavement was expected before earthwork starts. Earth moving equipment and hauling trucks would be moving in and out of the site during earthwork and excavation, creating struck-by hazards for the passing traffic and on-site workers. There would be trenching activities for the foundation and the APP needs to specify methods that can prevent cave-ins and protect the excavation faces. It is also possible that excavating at different depths creates fall hazards. Steel reinforcement and formwork for the foundation introduce cutting and abrasion hazards. The most expected hazard, unprotected reinforcement steel, is another major source of concern for the cast-in-place foundation concrete work. Concrete pumping operations in concert with the use of transit trucks present their own unique hazards such as exposure to concrete, unexpected pump hose kinking, and unexpected pressurized concrete. For FS39, the underground cistern structure, once being completed, would become a confined space and demand particular attention in the APP.

Masonry and structural steel form the superstructure throughout the FS39 project. It is not surprising that a significant amount of scaffold would be onsite for masonry construction, and fall hazards would be the most pressing concern for this phase of the project. Secondary to fall hazards are struck-by hazards created by tools and materials falling from heights. For the structural steelwork, fall hazards and struck-bys also apply and could be more severe. Because the fall hazards were above 10 feet, according to the WA L&I requirements, the project team would need to prepare a site-specific fall protection plan. This plan is different from an APP but can be used together with the APP to create a more seamless shield against potential hazards on-site. Figures 5.9, 5.10, 5.11, 5.12, and 5.13 exhibit a template fall protection work plan from the WA L&I for the construction industry. Contractors could use the template to examine in detail where the fall hazards are on a particular project and what can be done to mitigate or avoid the hazards.

FALL PROTECTION WORK PLAN – SAMPLE ONE

INSTRUCTIONS

A written fall protection work plan must be implemented by each employer on a job site where a fall hazard of 10 feet or greater exists, in accordance with Department of Labor and Industries, WISHA Regulations. **The plan must be specific for each work site.**
THIS WORK PLAN WILL BE AVAILABLE ON THE JOB SITE FOR INSPECTION.
Attached is a sample of a model fall protection work plan that may be filled out by each employer who has employees exposed above 10 feet. The following steps will help you fill out your plan.

1. FILL OUT THE SPECIFIC JOB INFORMATION.

Company Name: ____________________

Job Name: ____________________ Date: __________

Job Address: ____________________ City: __________

Job Foreman: ____________________ Jobsite Phone: __________

2. FALL HAZARDS IN THE WORK AREA

INCLUDE LOCATIONS AND DIMENSIONS FOR HAZARDS

Elevator shaft: __________ Stairwell: __________

Leading edge: __________ Window opening: __________

Outside static line: __________ Roof eave height: __________

Perimeter edge: __________ Roof perimeter dimensions: __________

Other fall hazards in the work area: ____________________

N-1

Figure 5.9 WA L&I fall protection work plan template—Part 1

3. METHOD OF FALL ARREST OR FALL RESTRAINT

(For fall protection equipment include details, such as manufacturer etc.)

Full body harness: ____________________ Body belt (Restraint only): ____________________

Lanyard: ____________________ Dropline: ____________________

Lifeline: ____________________ Restraint line: ____________________

Horizontal lifeline: ____________________ Rope grab: ____________________

Deceleration device: ____________________ Shock absorbing lanyard: ____________________

Locking snap hooks: ____________________ Safety nets: ____________________

Guard rails: ____________________ Anchorage points: ____________________

Catch platform: ____________________ Scaffolding platform: ____________________

Safety monitor: ____________________ Name of monitor, if used: ____________________

Other:

____

____

____

N-2

Figure 5.10 WA L&I fall protection work plan template—Part 2

For material handling, hoisting, and placement operations, the project team would use equipment such as a hydraulic lift and mobile crane. Hazards due to improper equipment maintenance or operations can be serious. Cutting and grinding concrete masonry unit (CMU) blocks can introduce silica dust and pose a health hazard. Lastly, the integrity and stability of the masonry walls or structural steel present a different type of concern, especially when strong local gusts or other weather-driven impacts occur.

Framing, sheathing, siding, and glazing would take place with the use of ladders and man lifts for the FS39's exterior enclosure. Improper use of ladders

4. ASSEMBLY, MAINTENANCE, INSPECTION, DISASSEMBLY PROCEDURE

Assembly and disassembly of all equipment will be done according to manufacturers' recommended procedures. (Include copies of manufacturer's data for each specific type of equipment used.)

Specific types of equipment on the job are:

____

____

____

____

____

____

____

A visual inspection of all safety equipment will be done daily or before each use, as stated in the Employee Training Packet. Any defective equipment will be tagged and removed from use immediately. The manufacturer's recommendations for maintenance and inspection will be followed.

____

____

____

____

5. HANDLING, STORAGE & SECURING OF TOOLS AND MATERIAL

Toe boards will be installed on all scaffolding to prevent tools and equipment from falling from scaffolding.

Other specific handling, storage and securing is as follows:

____

____

____

____

N-3

Figure 5.11 WA L&I fall protection work plan template—Part 3

or man lifts will result in severe fall hazards. Some weatherproofing work was expected around the rooftop during the summer season, creating heat stress in addition to the fall hazards for the workers who will access the rooftop.

Interior enclosure activities mainly included mechanical, electrical, and plumbing (MEP), interior framing, flooring, painting, finishing, and case work. Fall hazards would continue to surface with the use of ladders, rolling scaffolds, and stilts. Improper lighting or ventilation could cause concerns after the exterior enclosure is complete. The interference between the trades and spatial needs would reach its peak at this phase of the construction.

A condensed version of the FS39's APP developed by the authors is shown in Appendix B to demonstrate how project-specific hazards can be addressed

6. OVERHEAD PROTECTION

Hard hats are required on all job sites with the exception of those that have no exposure to overhead hazards. Warning signs will be posted to caution of existing hazards whenever they are present. In some cases, debris nets may be used if a condition warrants additional protection.

Additional overhead protection will include: ________________________________

____

____

____

Toe boards (at least 4 inches in height) will be installed along the edge of scaffolding and walking surfaces for a distance sufficient to protect employees below. Where tools, equipment or materials are piled higher than the top of the toe board, paneling or screening will be erected to protect employees below.

7. INJURED WORKER REMOVAL

Normal first aid procedures should be performed as the situation arises. If the area is safe for entry, the first aid should be done by a foreman or other certified individual.

Initiate Emergency Services – Dial 911 (where available)

Phone location: ________________________________
First aid location: ________________________________
Elevator location: ________________________________
Crane location: ________________________________
Other: __________________ Location: ____________________

Rescue considerations. When personal fall arrest systems are used, the employer must assure that employees can be promptly rescued or can rescue themselves should a fall occur. The availability of rescue personnel, ladders, or other rescue equipment should be evaluated. In some situations, equipment that allows employees to rescue themselves after the fall has been arrested may be desirable, such as devices that have descent capability.

Describe methods to be used for the removal of the injured worker(s):

N-4

Figure 5.12 WA L&I fall protection work plan template—Part 4

in an APP from a general contractor's perspective. The part "Phase Project Planning: Site-Specific Hazards" corresponds to most, if not all, of the FS39 hazards discussed in Section 5.3 whereas the technical sections in the APP identify the safe practices, rules, and precautions that field workers should use to mitigate or control the potential hazards. It should be noted that some of the technical sections such as Fall Protection, Scaffold Safety, Ladder Safety, PPE, Tool Safety, Hazards Communication, Fire Safety, Housekeeping, Burning and Welding, and Fuel Storage in Appendix B are common elements and can generally apply to every construction project.

____

____

____

____

8. TRAINING AND INSTRUCTION PROGRAM

All new employees will be given instructions on the proper use of fall protection devices before they begin work. They will sign a form stating they have been given this information. This form becomes part of the employee's personnel file.

The written fall protection work plan will be reviewed before work begins on the job site. Those employees attending will sign below. The fall protection equipment use will be reviewed regularly at the weekly safety meetings.

Date: ____________________

____________________	____________________
____________________	____________________
____________________	____________________
____________________	____________________
____________________	____________________
____________________	____________________
____________________	____________________
____________________	____________________

Foreman or Job Superintendent: ____________________

Prior to permitting employees into areas where fall hazards exist, all employees must be trained regarding fall protection work plan requirements. Inspection of fall protection devices/systems must be made to ensure compliance with WAC 296-155-24510.

N-5

Figure 5.13 WA L&I fall protection work plan template—Part 5

5.4 SITE MOBILIZATION

Site mobilization is when a contractor moves personnel, equipment, and operating supplies to the site in order to establish a site office and prepare for the operations at the site. Early development of a site logistics plan, such as the one in Figure 5.8, provides a well-thought-out plan that guides the site setup (e.g., perimeter of the site) as well as construction activities taking place throughout the entire project (e.g., material delivery, storage, and disposal). Typical site mobilization involves defining the administrative procedures (e.g., setting up meeting and reporting routines), acquiring tools, materials, and supplies, and installing temporary utilities, security gates/fencing, signage, and sanitary facilities. Due to the miscellaneous nature of site mobilization, a comprehensive checklist is often the best tool to help ensure that not a single item is left undone.

For FS39, perimeter and filter fabric fencings were established during site mobilization, as illustrated in Figure 5.14. Site fencing prevents the general public from accidentally entering the site and also provides protection against theft and vandalism of tools, equipment, and properties on-site. Filter fabric fencing was particularly necessary because the site was in a salmon watershed area, and the salmon are federally listed as endangered species. A temporary erosion and sedimentation collection system was required to ensure that sediment-laden water did not leave the site or enter the natural or public drainage system. The filter fabric fencing does so by filtering, redirecting, and impeding the flow of silt-laden water. Boundaries of the site perimeter fencing and filter fabric fencing can be found in the project's TESC/Site Demo Plan in Figure 5.15, where the dashed lines indicate the construction limits and the lines with small circles show the filter fabric fences. Existing power that had to be relocated by the City of Seattle before site clearing and grading is also indicated in Figure 5.15. Catch basin protections, as specified by the hexagon icons in Figure 5.15, also needed to be installed.

Figure 5.14 Perimeter and fabric filter fencings during project mobilization at FS39

(Courtesy of the Miller Hull Partnership, LLP.)

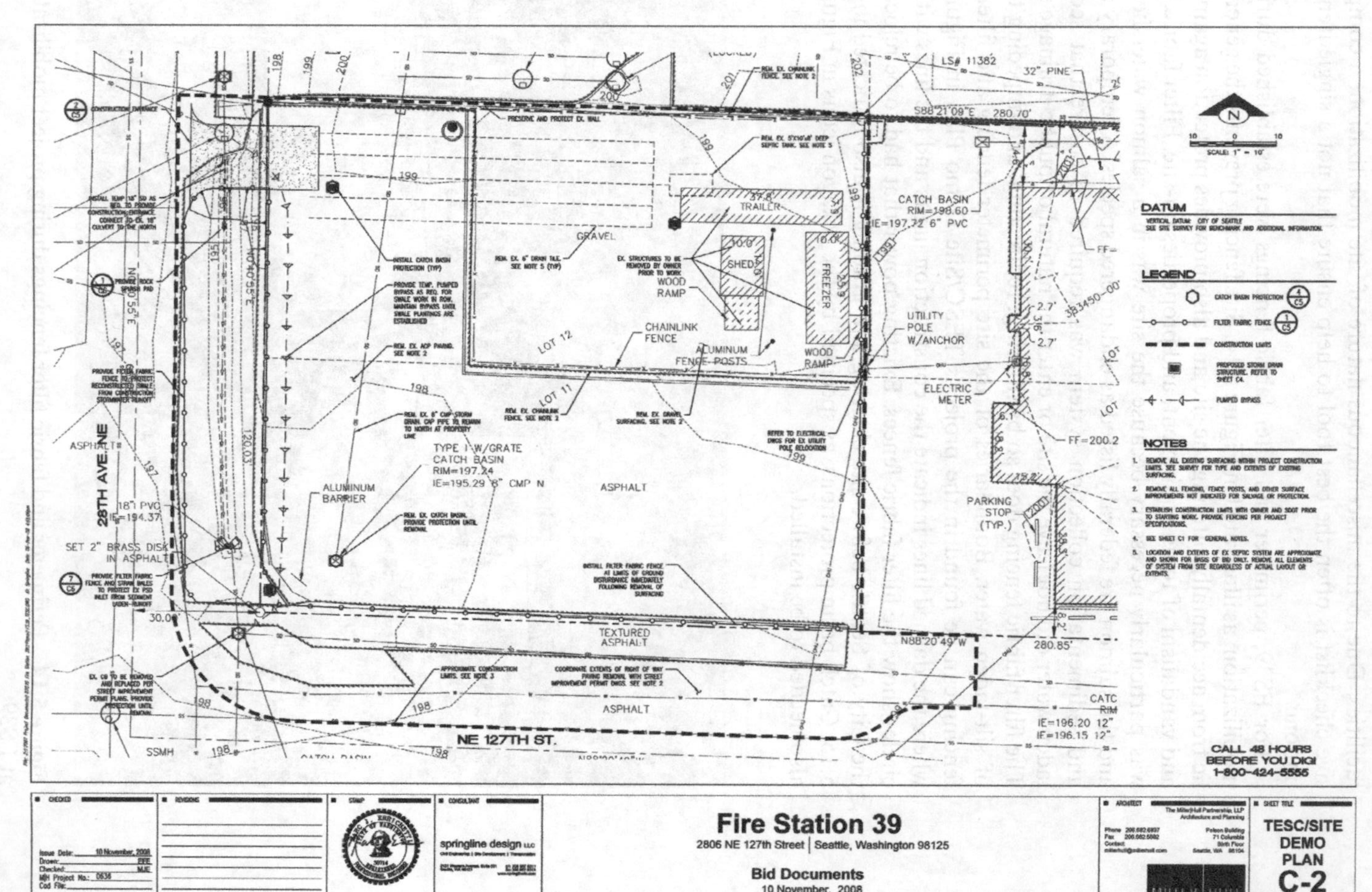

Figure 5.15 TESC and site demo plan for FS39

5.5 SUMMARY

This chapter introduced Fire Station 39 (FS39), which will be used throughout the book to illustrate the occupational hazards that are typical to the different phases involved in a construction project. Fire Station 39 is located in Lake City, Washington, and is a structural steel and masonry framed building. Images produced by the BIM exercise for the project are discussed in the chapter to illustrate the construction sequencing of FS39 and to foresee potential issues or logistical requirements of the site operations during preproject planning. Through the building information model, it is clear that the excavation and foundation work should begin at the south end of the site footprint in order to speed up the critical-path activities. The building information model also helps determine the site dynamics and informs the planning of site logistics such as the location of the tool shack and trailer office. After reviewing the FS39 building information model and discussing the site logistics, the chapter further looked into the project-specific safety and health hazards in the five major project phases (i.e., phased project planning). These phases are mobilization, earthwork and foundation, superstructure, exterior enclosure, and interior enclosure. Understanding project-specific hazards is imperative to the determination of safe practices, rules, and precautions that should be included and reflected by adequate technical sections in a project-specific APP. To accompany many of the key points discussed in the chapter, the authors have developed a condensed version of FS39's APP in Appendix B. Finally, the chapter discussed how a site can be set up during mobilization and provided examples from the FS39 project.

5.6 REVIEW QUESTIONS

1. What project characteristics should one consider during preproject planning?
2. How does BIM support preproject planning?
3. What are the key aspects that a site logistics plan should entail?
4. What rationale was used to define a site logistics plan for FS39?
5. Does the rationale identified in Question 4 apply to other projects as well? Why or why not?
6. How does phased project planning inform a project's site-specific APP?
7. What mobilization activities were involved in the FS39 project?
8. Do the activities identified in Question 7 apply to other projects as well? Why or why not?
9. What is fabric filter fencing and why is it used during project mobilization?

5.7 EXERCISES

1. Research the FS39 project and discuss project characteristics that have not been identified but are crucial to the preproject planning activity.
2. Find a construction company in your community and interview a site superintendent about his or her current project. What are the site logistics of the project? What are the rationales behind these logistics? Do these rationales apply to other projects?
3. Find a construction company in your community and interview a site superintendent about his or her current project. What are the major hazards involved in each project phase? Ask the superintendent how his or her current project's APP (if there is one) addresses these hazards.
4. What does stormwater control mean and why is it considered during project mobilization?

6

SAFETY FOR SITE PREPARATION, EARTHWORK, AND FOUNDATION

6.1 INTRODUCTION

Once the construction contractor had mobilized and occupied the project site, the initial construction tasks were to install temporary fencing, install stormwater runoff control measures, and relocate electrical service to the site. At the same time, crews were used to remove existing pavement, a fence, and an underground septic tank and connected drain field. Once the site was cleared, excavation for the foundation was initiated and site utilities installed. Next the building footings and foundation walls were constructed, and the slab foundation was placed.

This chapter addresses some of the risks construction workers experienced during this phase of the project and measures taken to minimize the potential for injury. OSHA standards which are relevant to the hazards discussed in this chapter are also recapped in Appendix E. The first step is to plan the work to be performed during this phase of the construction. As discussed in Chapter 5, the building information model can be used to visualize the work to be performed after the site has been cleared. Figure 6.1 shows the location of the footings to be constructed, Figure 6.2 shows the location of the cistern, which required the deepest excavation on this project, and Figure 6.3 shows the location of the stem walls.

6.2 SITE PREPARATION

The site selected for construction of FS39 contained an asphalt parking lot, a chain link fence enclosed area containing three temporary structures, and a septic tank with tile drain field. The project owner removed the temporary structures, but the construction contractor was responsible for removing the fence, the asphalt pavement, the septic tank, the tile drain field, and the existing overhead electrical service to the site.

The chain link fence was removed from the existing posts, and the posts were then pulled out. The septic tank and tile drain field were dug out by a hydraulic excavator and removed from the site. The excavator was also used to remove the asphalt pavement, as shown in Figure 6.4. A good traffic control plan is needed

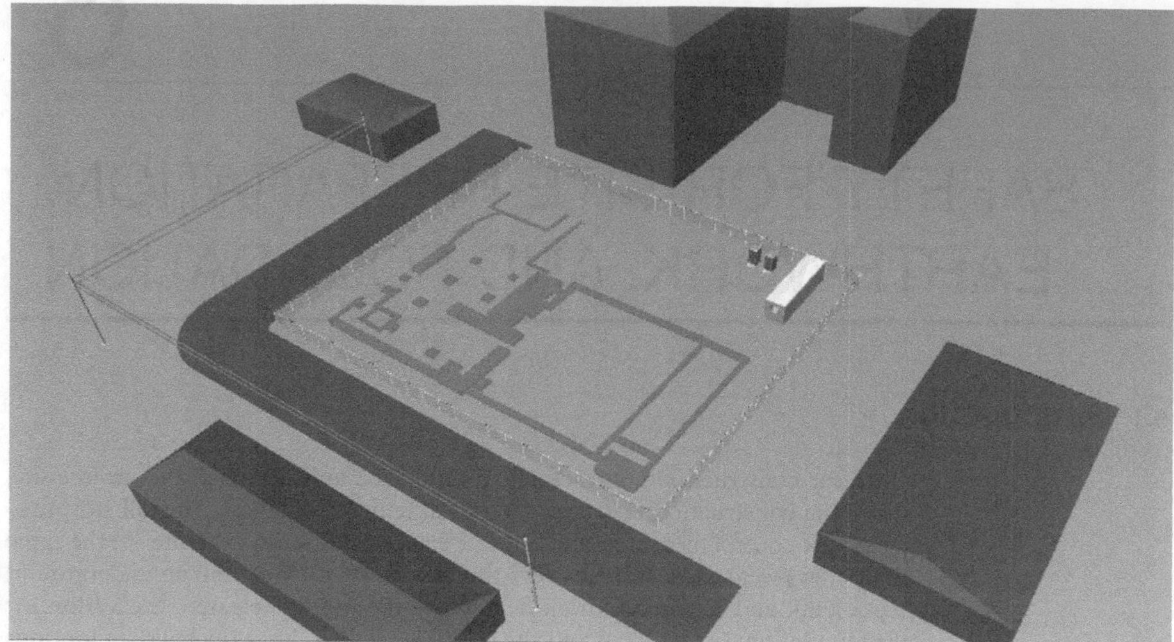

Figure 6.1 Building information model showing footing locations

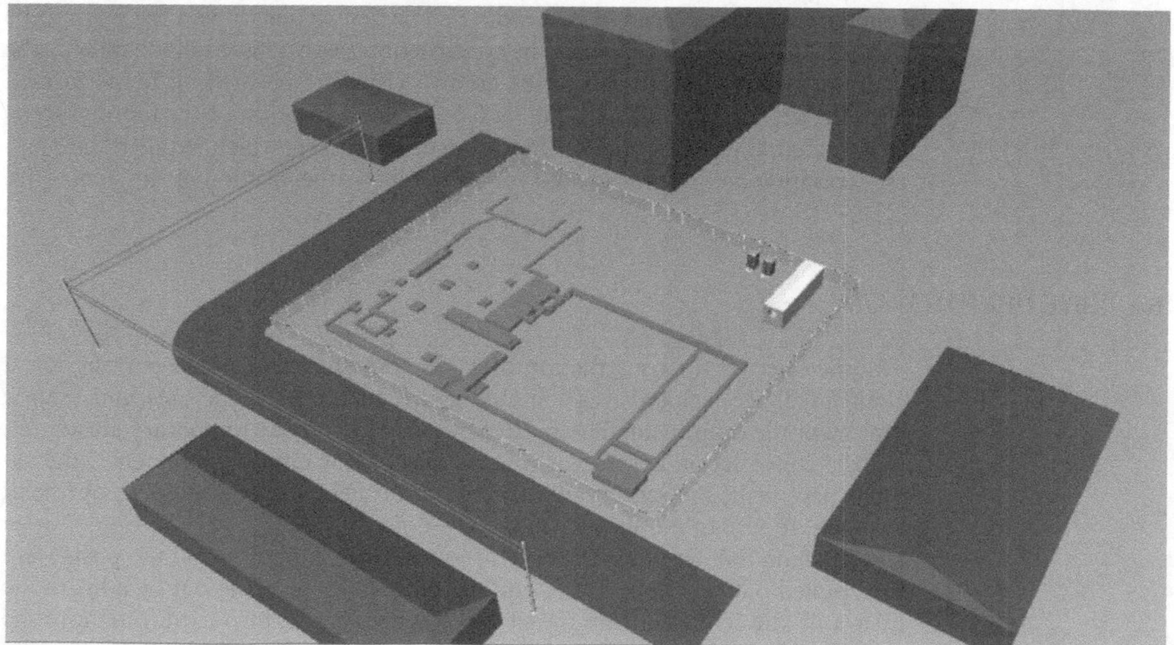

Figure 6.2 Building information model showing cistern location

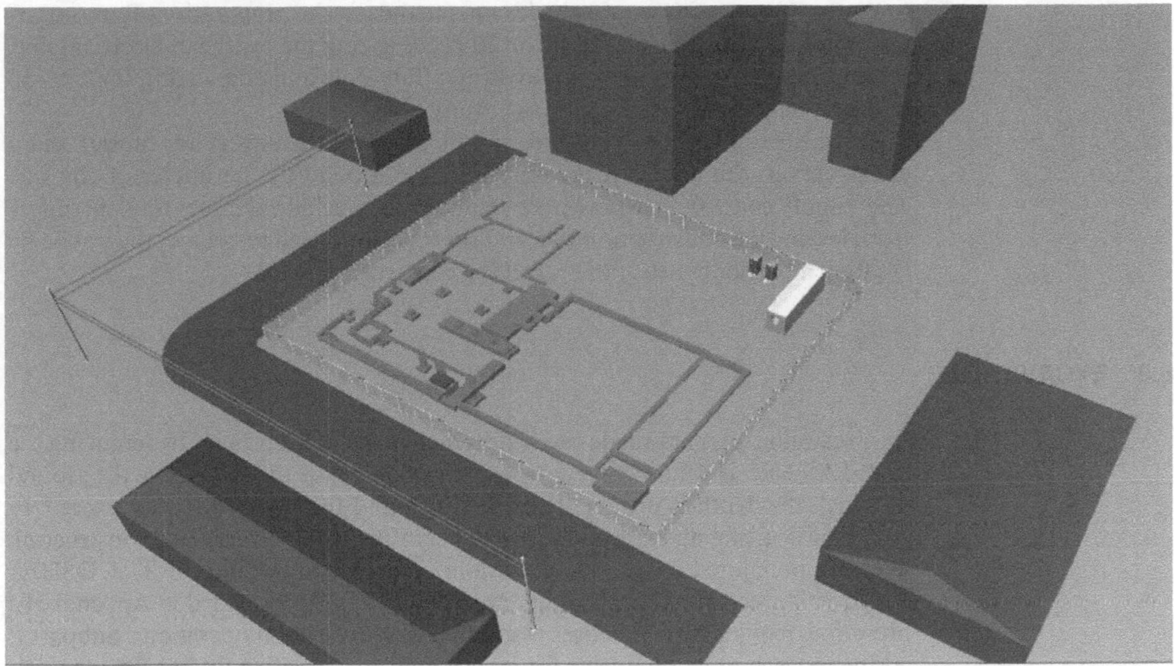

Figure 6.3 Building information model showing stem wall locations

Figure 6.4 Clearing the construction site and removing existing asphalt

(Courtesy of The Miller Hull Partnership, LLP.)

to control the movement of vehicles on-site during site preparation. Removal of the overhead electrical service involved deenergizing the overhead electrical line, disconnecting the wires from an existing utility pole and connecting them to an adjacent building, and removing the utility pole.

Once the site had been cleared, runoff control measures (filter fabric) and a construction (chain link) fence were installed around the perimeter of the site. The runoff control measures were required to preclude sediment-laden runoff from leaving the construction site, and the fence was used to protect the public by creating a barrier between them and the construction work.

6.3 EXCAVATION

An excavation is a man-made cavity or trench made in the ground by removing the earth. A trench is defined as a narrow excavation below the surface of the ground in which the depth is greater than the width and the width does not exceed 15 feet. Constructing an excavation can be very dangerous. Every year construction workers are injured or die due to cave-in of excavations. Subpart P of OSHA's construction safety and health standard (29 CFR 1926) (provided in Appendix E) prescribes required safe practices associated with excavation operations. Subpart P addresses general requirements for excavations and specific requirements for preventing cave-ins and other related hazards.

Many on-the-job accidents are the result of inadequate planning. Correcting mistakes in sloping or shoring an excavation once work is underway usually results in delay and extra cost. Excavations normally have one or more sides and are open for more than 24 hours. Anyone working on an excavation must be protected from the hazard of cave-in by an adequate protective system that meets OSHA requirements. Three systems are allowed:

- Sloping the soil away from the edge of the excavation
- Shoring the sides of the excavation
- Using a shielding system to protect the workers

The maximum permitted slope is a function of the soil type. OSHA adopted the textural soil classification system from the U.S. Department of Agriculture (USDA), which classifies a soil's type based on the percentage of sand, clay, and silt that is in the soil sample. Figure 6.5 shows the USDA textural soil classification system. There are three types of soil: A, B, and C. The higher the percentage of clay (toward the upper part of the classification triangle), the higher the chance that the soil will be considered type A. On the other hand, a high concentration of sand (toward the lower left corner of the classification triangle) is typical for the type C soil and will cause the soil to shift and pour onto the workers.

The sloping (run: rise) requirements from OSHA for the three types of soil are:

- *Type A* soil is a cohesive soil that is usually composed of clay, silty clay, sandy clay, or loam. It has a maximum permitted slope of 0.75 to 1.

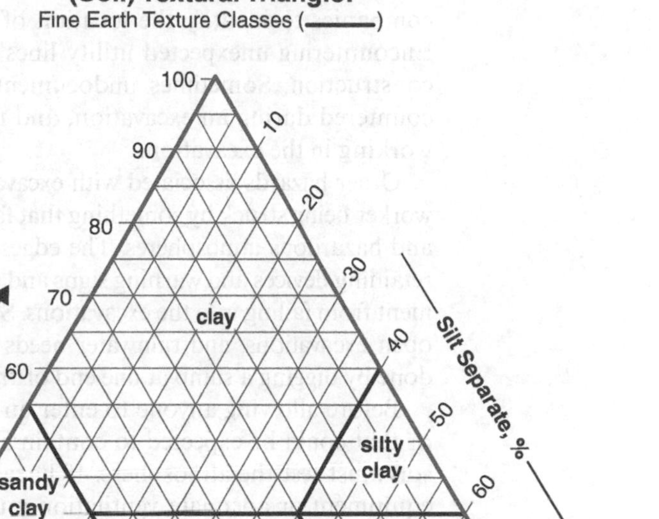
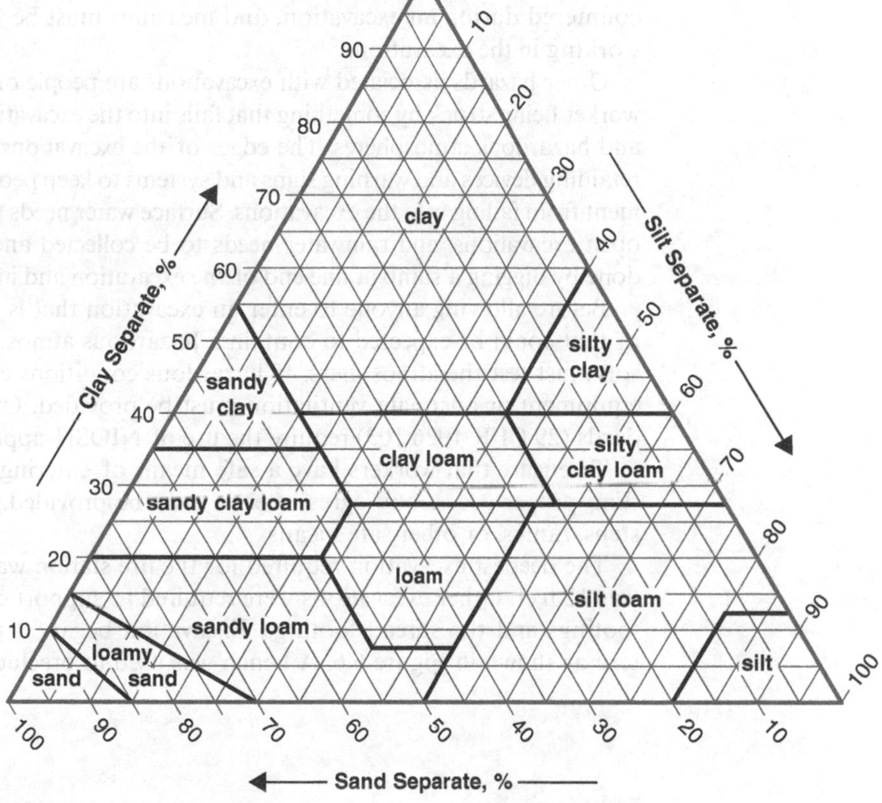

Figure 6.5 USDA textural soil classification system

- *Type B* soil is a granular cohesionless soil, including angular gravel, crushed rock, loam, or silty loam. It has a maximum permitted slope of 1 to 1.
- *Type C* soil includes gravel, sand, and loamy sand. It has a maximum slope of 1.5 to 1.

Shoring systems may be used to support the sides of an open excavation. This may involve driving vertical H-piles and installing horizontal timbers to support the side of the excavation. Another approach is to install metal shoring systems. For trenches, many contractors choose to use trench boxes which protect workers from the hazard of cave-in.

OSHA requires that all excavations be inspected daily by a competent person prior to any workers entering the excavation. The competent person inspects for any changes that have occurred in the excavation due to ground vibration or rainfall. The competent person must have the knowledge to be able to recognize any hazard present and the authority to direct appropriate corrective action.

Prior to initiating any excavation, the contractor must contact local utility companies to identify the location of any utility lines crossing the excavation. Encountering unexpected utility lines can be a significant hazard in excavation construction. Sometimes undocumented utilities and other structures are encountered during an excavation, and measures must be taken to protect people working in the excavation.

Other hazards associated with excavations are people or equipment falling in, a worker being struck by something that falls into the excavation, water accumulation, and hazardous atmospheres. The edges of the excavations must be protected with retaining devices and warning signs and systems to keep people, material, and equipment from falling into the excavations. Surface water needs to be diverted away from open excavations, and rainwater needs to be collected and removed. This can be done by digging a sump at one end of the excavation and installing a suction pump.

Before allowing anyone to enter an excavation that is 4 feet or more in depth or that could be expected to contain a hazardous atmosphere, a competent person must test the atmosphere. If hazardous conditions exist, proper respiratory equipment or adequate ventilation must be provided. OSHA's respiration standards (29 CFR 1926.103) require the use of NIOSH-approved respirators.

To ensure that workers have a safe means of entering and exiting an excavation, proper access and egress means must be provided. These may be ladders, steps, ramps, or other safe means.

The deepest excavation required for the fire station was for the cistern, which was 12 feet. Other excavations were required to support construction of the wall footings and the spread footings. Excavation began at the location of the cistern as shown in Figure 6.6. A bench was used to preclude the need for shoring.

Figure 6.6 Excavation for cistern

(Courtesy of The Miller Hull Partnership, LLP.)

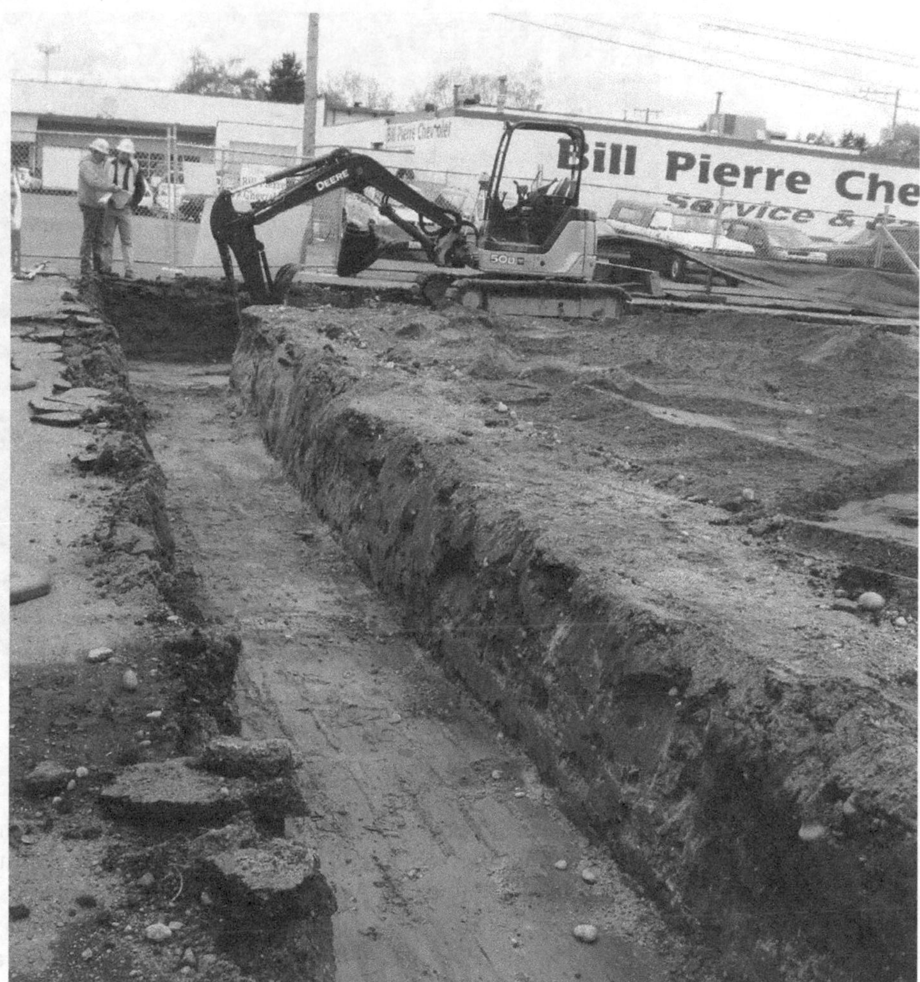

Figure 6.7 Excavation for wall footings
(Courtesy of The Miller Hull Partnership, LLP.)

Warning tape and barricades were installed to warn workers of the open excavation to minimize the potential for falling. A small excavator was used to excavate trenches for the wall footings, as shown in Figure 6.7.

6.4 FOUNDATION CONSTRUCTION

A typical concrete foundation starts with the construction of the concrete forms and installation of the reinforcing steel. Once the steel is in place, the concrete may be placed using a concrete pump or concrete bucket or placed directly from a concrete transit truck. Concrete formwork must be designed and constructed

Figure 6.8 Footing formwork construction

(Courtesy of The Miller Hull Partnership, LLP.)

so that it can support safely, both vertically and laterally, the load of the concrete placed inside. Workers employed in concrete placement and finishing must wear protective clothing to preclude contact with the concrete.

Figure 6.8 shows two workers constructing the wood forms for one of the mat footings for the fire station. The bricks shown in the figure are being used to keep the horizontal reinforcing steel off of the ground to ensure that the bars are fully encased within the completed concrete footing. Vertical reinforcing steel bars are installed within the forms to tie the foundation elements to the rest of the structure.

Once the forms have been constructed and all reinforcing steel installed, concrete is placed within the forms. Reinforcing steel that protrudes above the concrete must be capped with protective caps to preclude worker injury. These caps are shown in Figure 6.9. OSHA requirements for installing these caps will be discussed in more detail in Chapter 7.

Wood concrete forms were constructed for the cistern, and reinforcing steel bars were installed. Concrete was placed within the forms and vibrated to ensure consolidation. Once the concrete had cured, the forms were removed. Because the cistern excavation was over 4 feet, it is considered a confined space, requiring daily testing of the air quality and constant monitoring of people working in the space. This hazard was identified in the APP shown in Appendix B. The most common hazard associated with such spaces is lack of oxygen, and anyone working in such spaces requires special training.

Figure 6.9 Concrete footings

(Courtesy of The Miller Hull Partnership, LLP.)

6.5 SITE UTILITIES

Site utility construction involved the installation of both water and sewer connections to the existing main lines, construction of an underground stormwater detention tank, and connecting the electrical service lines to an adjacent building in lieu of installing a new utility pole. Figure 6.10 shows FS39's utility plan and

Figure 6.10 Fire Station 39's utility plan

Figure 6.11 Fire Station 39's utility piping placement at the southwest corner of the site

(Courtesy of The Miller Hull Partnership, LLP.)

indicates locations of the required utility mains or connections. Trenches were excavated and pipes were installed for both the water and sewer connections. An excavation was made to facilitate the construction of the underground detention tank. The tank was installed and then backfilled and graded to the slope requirements shown on the site plans. Since a new utility pole was not required, the electrical service was moved to the adjacent building at the same time as the existing service was removed. Figure 6.11 shows the utility piping placement at the southwest corner of the fire station.

6.6 OTHER SITE WORK

Other site work involved in the fire station project included:

- Backfilling the foundation
- Grading the site to conform to the slopes specified in the construction drawings
- Forming and placement of sidewalks
- Constructing the planting areas
- Placing concrete pavement
- Placing recycled cobblestones provided by the project owner

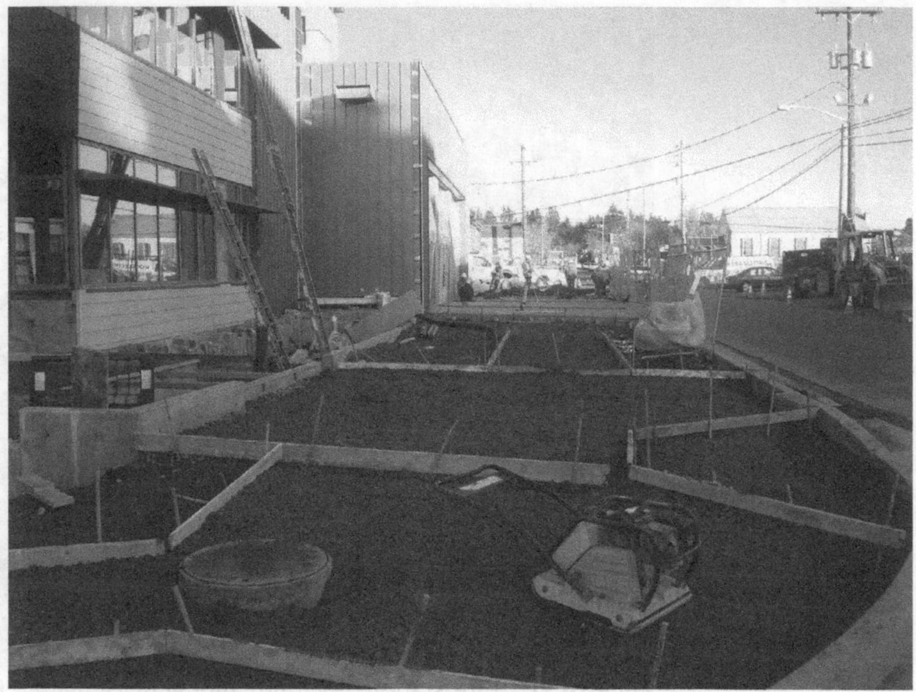

Figure 6.12 Formwork for sidewalk and planter areas

(Courtesy of The Miller Hull Partnership, LLP.)

Any mechanized equipment used on-site must conform to Subpart O of OSHA's construction safety and health standard (29 CFR 1926). The most common hazards relate to backing equipment and equipment operators not seeing other workers. The use of back-up alarms or ground guides and wearing high-visibility clothing help mitigate these risks.

Once the site grading was complete, the formwork for the sidewalks and planting areas was constructed, as shown in Figure 6.12.

6.7 SUMMARY

The major safety hazards associated with site preparation, excavation, and foundation are:

- Control of vehicles and equipment on-site
- Excavations
- Confined space construction
- Utility relocation

A good traffic control plan is essential to control vehicles entering and leaving the site safely. Materials removed from the site need to be properly controlled and disposed of. Stormwater needs to be controlled to comply with environmental regulations.

- Excavation procedures need to include systems to protect against cave-in. Approved systems include:
- Sloping the soil
- Shoring the sides
- Using a shielding system

All excavations must be inspected daily by a competent person to ensure conditions have not changed. Barricades and signs need to be installed to keep people or equipment from falling into an excavation. Confined spaces deeper than 4 feet need to be inspected daily for adequate oxygen, and workers entering such spaces need special training to understand the risk faced while working in such spaces.

6.8 REVIEW QUESTIONS

1. How can the building information model be used to determine potential hazards during site preparation and excavation?
2. What factors should be considered in developing a traffic control plan for a project site?
3. What are three systems that are allowed in excavation construction to protect against cave-in?
4. What is the difference between type A and type B soil?
5. What is a competent person?
6. Besides cave-in, what are other hazards associated with excavation construction?
7. What is a trench box, and how is it used in trench construction?
8. What process should be followed to locate existing utility lines?
9. What actions must be taken prior to allowing someone to enter a confined space?
10. What hazards are faced by workers constructing concrete forms and installing reinforcing steel?

6.9 EXERCISES

1. Research three shoring systems that can be used in excavation construction and determine the advantages and disadvantages of each system.
2. Research the types of respirators that are available for use when working in a confined space.
3. Research the process that must be followed if contaminated soil is encountered during site excavation.
4. Determine the training that a foreman must complete prior to being considered a competent person.

7

SAFETY FOR THE SUPERSTRUCTURE

7.1 INTRODUCTION

After the completion of site excavation and foundation work, a project moves into the phase of superstructure construction. Regardless of the types of superstructures involved, the phase is typically featured by fall hazards and the hazards of not maintaining the integrity of the superstructure during construction. Furthermore, heavy equipment is used to hoist materials into place, creating the hazards of falling objects. While the superstructure is being constructed and raised, the exposure to overhead utilities also increases. It is a phase when a project gets into its peak production and at the same time introduces three out of the four common types of hazard in construction, including "fall from height," "struck-by," and "electrocution." Together with "caught in or between," these four hazards are often referred to as the OSHA Focus Four.

This chapter looks into masonry construction and steel erection work at the FS39 project with an emphasis on three out of the OSHA Focus Four hazards. The chapter discusses the structural construction activities that take place, hazards associated with the activities, subparts of OSHA safety standards that regulate the activities, and safe working tips that help minimize the hazards. Images from the building information model introduced in Chapter 5 are illustrated in the chapter to highlight project-specific conditions that should be taken into consideration when planning for safety. Finally, the chapter touches upon the safety issues during floor construction. As in the case of FS39, slab on grade (SOG) was best scheduled after steel erection was completed in order to minimize the overhead hazards during steel construction. OSHA standards which are relevant to the topics discussed in this chapter can be found in Appendix E.

7.2 MASONRY CONSTRUCTION

Masonry application is well suited for many types of building construction because of its versatility, durability, and low maintenance. For FS39, the design firm had selected reinforced CMUs as the main method to construct the walls for the apparatus bay, the storage wing, and the fueling shelter located at the northwest corner of the site. Figures 7.1 and 7.2 are images produced by the building information model introduced in Chapter 5. The images help highlight the potential fall hazards associated with the wall construction, especially near the hose drying tower that extends beyond the roof top. The project's foundation plan S2.0 as illustrated in Figure 7.3 indicates the locations of these masonry structures and also suggests that most of the masonry walls sit on top of the wall footings. As Section 5.2 discusses, it would make the most sense to start the excavation at the cistern and apparatus bay stem walls so that their concrete foundation could be completed as soon as possible to kick start masonry construction.

The project's foundation section detail plan S5.1 as illustrated in Figure 7.4 specifies the reinforcement requirements for connecting the footings, SOG, and masonry walls. Additional vertical and horizontal reinforcement is also typical

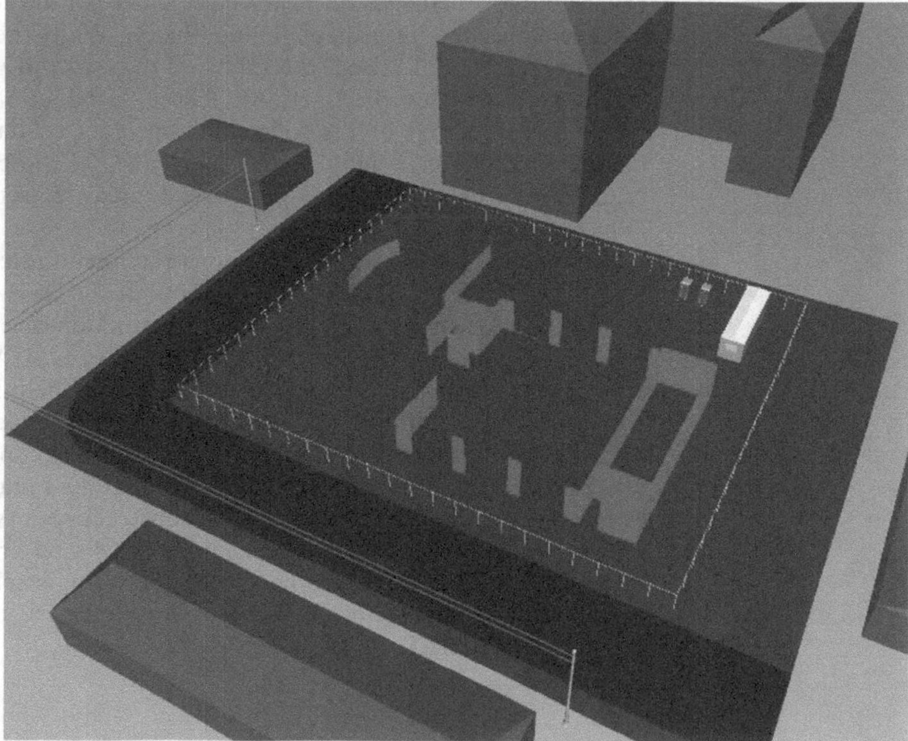

Figure 7.1 Building information model showing the first level of CMU

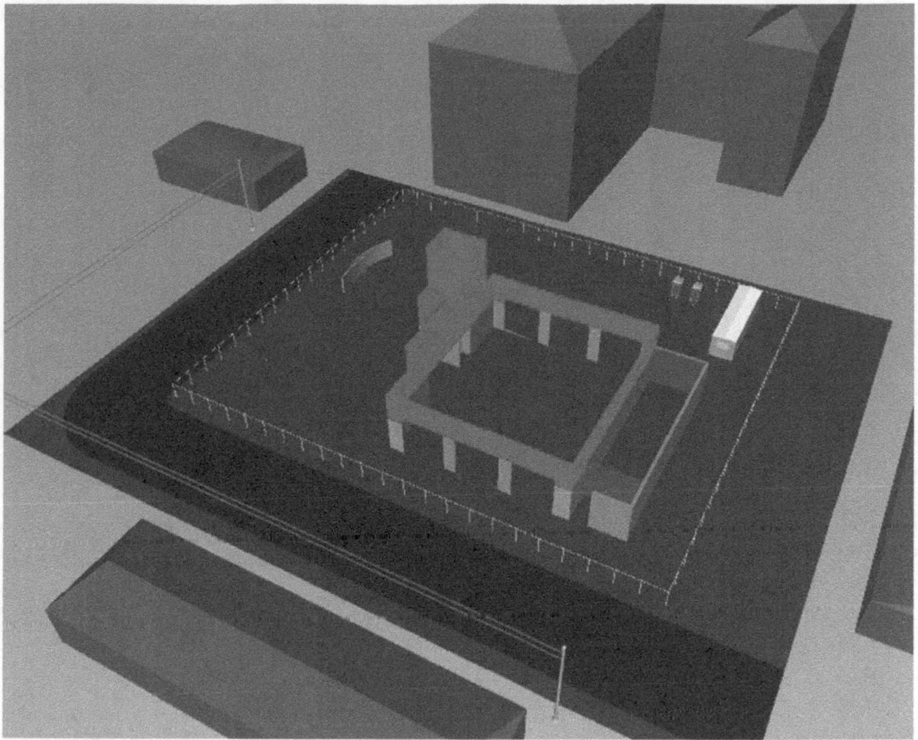

Figure 7.2 Building information model showing the second and third levels of CMU

to provide structural strength and to prevent cracking of the CMU walls. Not surprisingly, protruding reinforcement presents an immediate hazard that could cause serious impalement for the employers on-site even before masonry construction starts. OSHA standard 29 CFR 1926.701(b) requires that "All protruding reinforcing steel, onto and into which employees could fall, shall be guarded to eliminate the hazard of impalement." The use of rebar caps, such as the ones in Figure 7.5, is a fairly common solution to work around protruding reinforcement. It is worth mentioning that rebar caps come in different shapes and designs (steel-reinforced square covers versus mushroom-style plastic covers) and are intended for various applications. Only rebar caps that are designed to provide proper protection for the intended application should be used. Quite often some rebar caps are knocked off from the reinforcement without catching anyone's attention. Having a safe working culture on-site will encourage the employees who spot the missing caps to attend to the problem and take an active role in the safe operations in the field. Routine job walks completed by the site supervisors will also help rectify the problem. If adequate rebar caps are not accessible or available, 2 × 4 wooden troughs are accepted in some states as an alternative method.

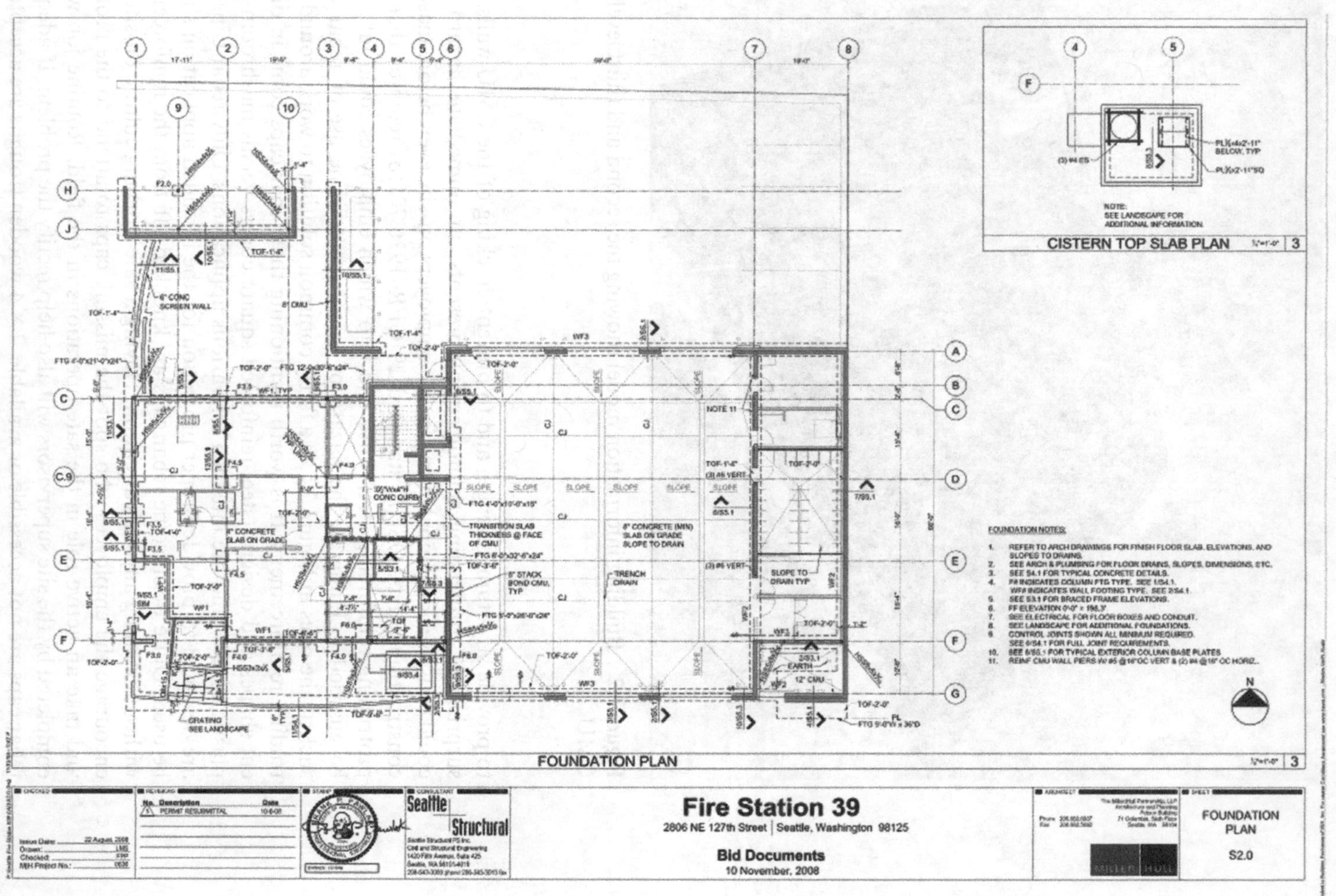

Figure 7.3 FS39's foundation plan S2.0 showing the masonry wall locations

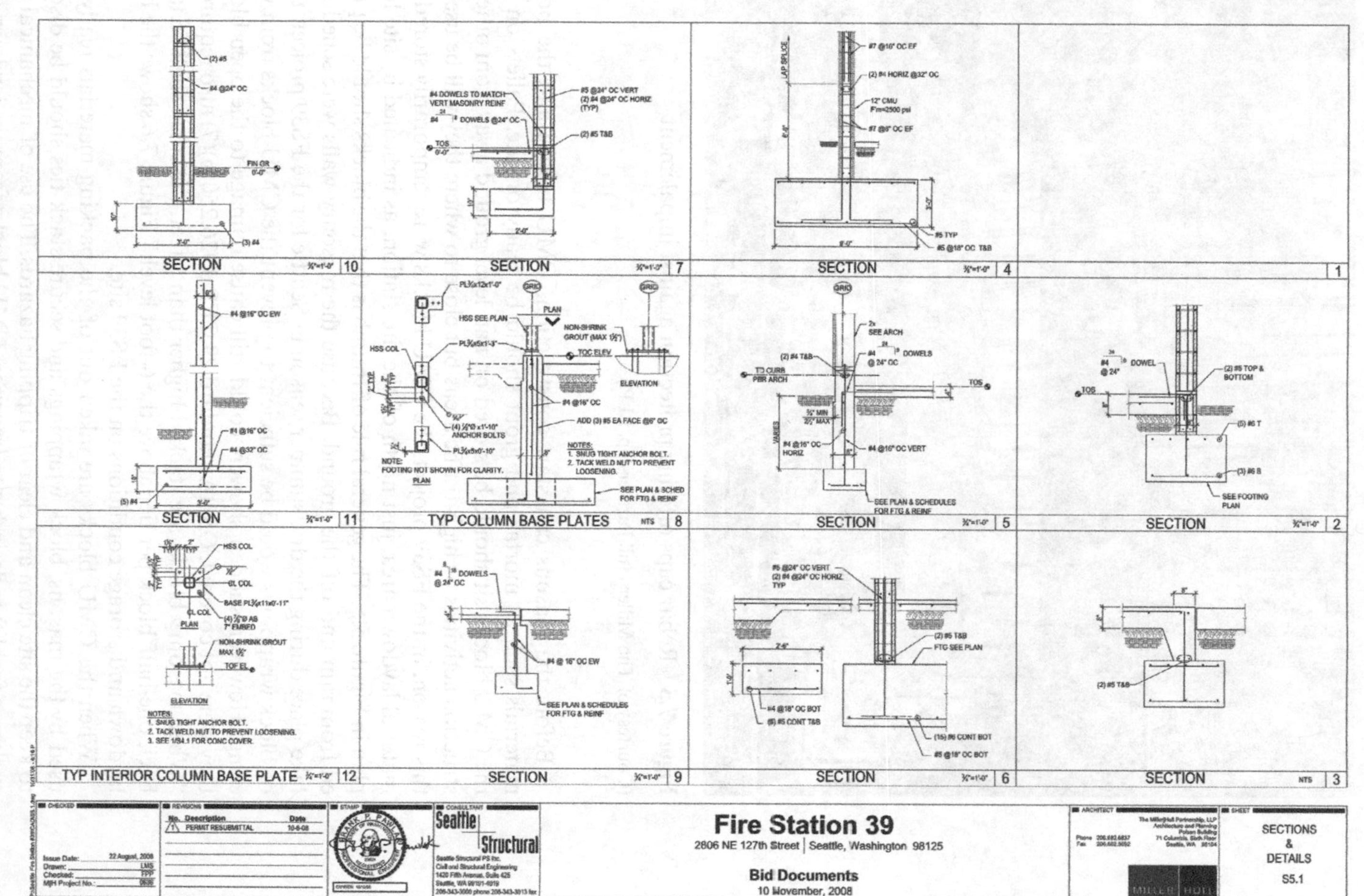

Figure 7.4 FS39's foundation detail plan S5.1 showing how masonry walls are connected to the wall footings

Figure 7.5 Rebar caps offering protection against impalement

(Courtesy of The Miller Hull Partnership, LLP.)

Before the masons can start construction, CMU blocks and the bonding materials such as mortar and grout should be ready on-site. Pallets on which the CMU blocks sit should be stored on flat, level ground away from other construction activities or high-traffic areas but close to where they will be used. For this reason, in the FS39 project, the CMU blocks were temporarily stored at the material laydown area, just north of the fire station, as indicated by site logistic plan in Figure 5.5. The general rule of thumb is that the blocks should be protected from rain and off the ground. Because the masonry walls were scheduled to take place during the dry, summer season in Seattle for the FS39 project, the use of block wrappings would be sufficient to prevent the CMU blocks from weathering. However, sudden showers could still cause damage to the open blocks if they were not covered. OSHA standard 29 CFR 1926.250(b)(7) also requires that "When masonry blocks are stacked higher than 6 feet, the stack shall be tapered back one-half block per tier above the 6-foot level." Figure 7.6 shows the blocks' laydown and storage conditions on the FS39 site.

When the CMU blocks are taken out of the packing materials ready to be used by the masons, block wrappings and secure stack ties should be discarded to keep the site clean and clear of tripping hazards. The use of mechanical means such as a forklift is desirable to transport CMU blocks from the laydown area to the scaffold landing area where the blocks are loaded for laying and placement.

Figure 7.6 On-site storage and protection of the CMU blocks
(Courtesy of The Miller Hull Partnership, LLP.)

Figure 7.7 shows the use of a forklift for handling CMU blocks at FS39. Because of the heavy loads of CMU blocks, excessive stress and strain associated with the manual, repetitive handling of the blocks can occur, especially considering that medium-weight blocks are specified for the masonry wall structures in the FS39 project. For this reason, adequate block stacking and staging heights are recommended to help reduce the need to lift CMU blocks below knees or above shoulders. In addition, dropped blocks could cause slips, trips, and struck-by hazards and should be cleaned up soon.

Forklifts are commonly seen on construction sites and are considered power industrial trucks (PITs), covered by OSHA's general industry standards 29 CFR 1019.178. The biggest cause of fatal forklift accident is the overturned forklift and it is important that forklift operators keep the load low to the ground, watch the center of balance of the load, and be mindful of shifting loads. Operators should not add weight to the back of a forklift to increase the load capacity and should always wear seat belts in case an overturn occurs. Forklifts are also frequently associated with near-miss events in which field employees are nearly ran over by these powerful vehicles. General contractors should establish a policy that limits how fast a forklift can be operated in the field. Pedestrians always have the right of way and having a good habit of sounding the horn to warn employees at intersections will go a long way.

Figure 7.7 Mechanical means such as the use of a forklift are preferred when handling masonry blocks on-site

(Courtesy of The Miller Hull Partnership, LLP.)

During masonry construction, block cutting and getting into contact with the bonding chemicals introduce health hazards that call for the use of particular PPE, including safety boots, safety gloves, safety glasses, respirators, and hearing protection. The use of PPE provides a second tier of safety measure after engineering controls such as using dust suppression or extraction systems or administrative controls like rotating workers for the task of saw cutting have been applied. Safety boots prevent masonry workers from having foot injuries in the event of dropped CMU blocks whereas safety gloves guard the workers against sharp edges and direct contact with cement. Safety glasses protect workers from flying masonry pieces and respirators help filter out the silica dusts and purify the breath-in air during saw cutting. Typical noises found on construction sites due to the use of masonry saws are at 95 decibels (dB) or more. OSHA requires that ear protective devices be provided and used when the noise exposure is at or exceeds the levels defined in 29 CFR 1926.52(a), Table D-2. For the reader's convenience, information from the table is provided in Table 7.1. In the case of FS39, because the site was under Washington State's jurisdiction, the rules on hearing protection are even more stringent (required when the noise level is at or exceeds 85 dB with 8 hours exposure). Masonry saws themselves present additional hazards and should be guarded with a semicircular enclosure. Since most

Table 7.1 OSHA Permissible Noise Exposures

Duration per Day, hours	Sound Level dBA Slow Response
8	90
6	92
4	95
3	97
2	100
1 and ½	102
1	105
½	110
¼ or less	115

of the masonry work will be completed with the use of scaffolding, besides the previously discussed PPE, hard hats for head protection should also be used to avoid overhead, struck-by hazards coming from falling materials or tools.

The project's foundation structural drawing S3.2 in Figure 7.8 shows the general elevation of the CMU walls, and based on the architectural drawings, the highest elevation of the CMU walls was just a little over 39 feet. The height of the masonry walls created the overturn hazard, especially when local wind gust picks up. In the northwest region, where Seattle is located, the wind gust can be as high as 50 mph. Therefore, based on the OSHA standards 29 CFR 1926.706, a limited-access zone should be established on the unscaffolded side before the wall construction starts. This would limit the zone access to only the employees who are actively engaged in constructing the wall. Once the wall reached 8 feet in height, it needed to be braced until permanent supporting elements of the structure are in place. Figure 7.9 shows how a bracing was attached to the doorjamb in order to ensure the stability of the CMU wall on FS39.

The height of the masonry walls also implied that a good amount of scaffolding was expected when constructing the apparatus bay masonry walls. There are many hazards associated with the use of scaffolds, including falls, being struck by a falling object, scaffold collapse, and electrocution. On FS39, the scaffolding used was fabricated tubular welded frame scaffold, as illustrated in Figure 7.10. This particular type of scaffold system consists of a platform supported on fabricated end frames with integral posts, horizontal bearers, and intermediate members. Level and sound footing at the base, proper scaffold access, adequate structure and platform composition, upright stability, and appropriate fall protection constitute the minimum safety requirements when working with tubular welded frame scaffolds.

To set up a level and sound footing at the base for tubular welded frame scaffold, contractors can use base plats or mud sills. Using cardboard, scrap wood, or cement blocks creates poor scaffold foundation and is not acceptable. Workers should

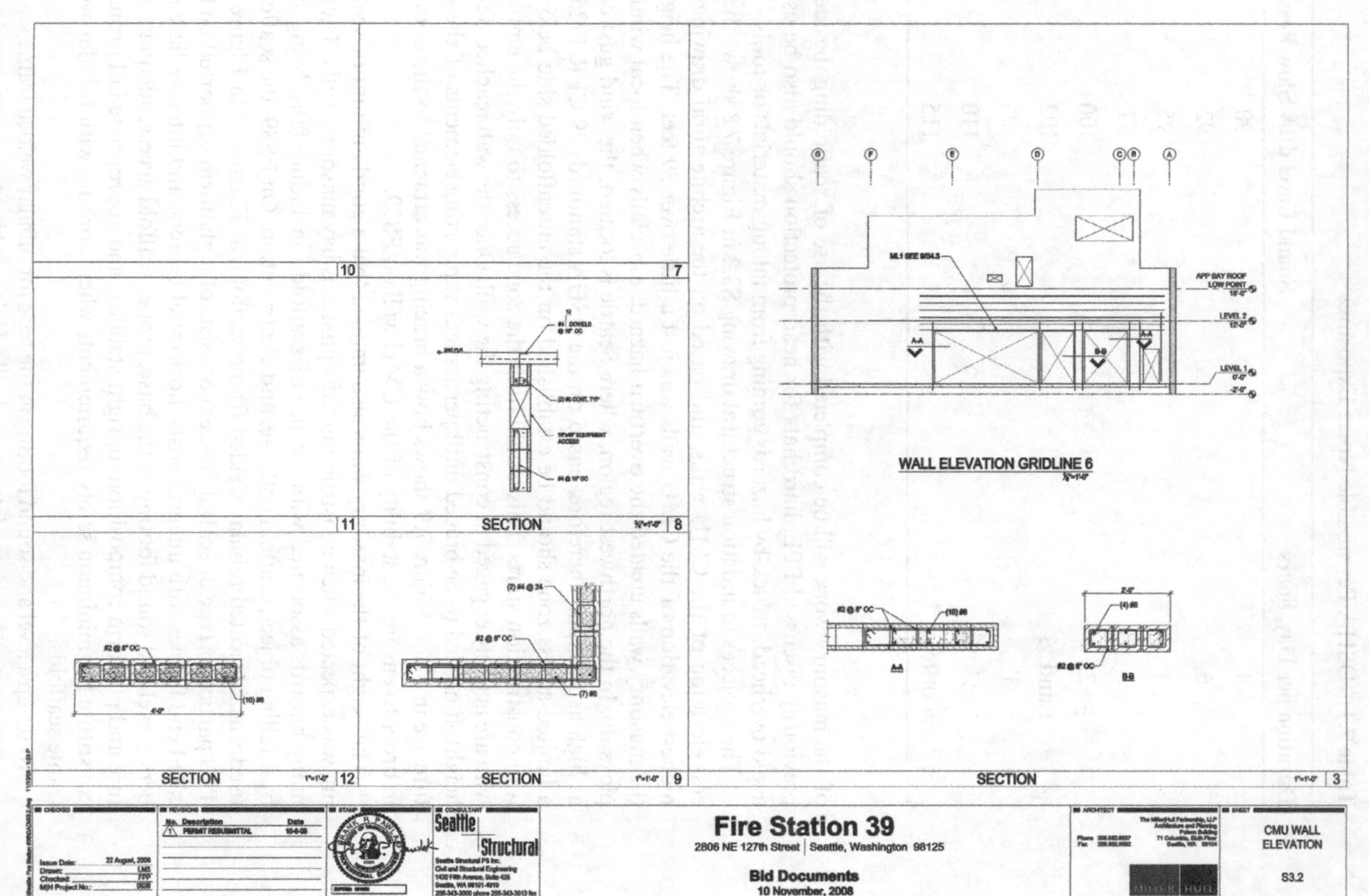

Figure 7.8 FS39's foundation structural drawing S3.2 illustrating the CMU wall elevations

Figure 7.9 Bracing masonry walls

(Courtesy of The Miller Hull Partnership, LLP.)

Figure 7.10 Fabricated tubular welded frame scaffold

(Courtesy of The Miller Hull Partnership, LLP.)

be trained to use proper access (e.g., ladder in the center in Figure 7.10) instead of climbing the cross bracing to reach higher stages of the scaffold. OSHA standards 29 CFR 1926.451 (a) and (b) describe the load capacity and platform construction for acceptable scaffolding. Each scaffold and scaffold component shall be capable of supporting, without failure, its own weight and at least four times the maximum intended load applied or transmitted to it. The platform should be fully planked (or decked) with sufficient width and be close enough to the working surface. The scaffold should also be properly anchored to the masonry structure to ensure its structural stability according to OSHA standard 29 CFR 1926.451(c), "Criteria for supported scaffold." The erection, movement, dismantlement, or alteration of a scaffold has to be under the supervision of a competent person.

Fall protection consists of either personal fall arrest systems (PFASs) or guardrail systems. According to OSHA standard 29 CFR 1926.451(g), "Fall protection," scaffolds more than 10 feet above a lower level shall be protected from falling to that lower level. PFASs arrest a falling employee whereas a guardrail system restrains an employee from falling. However, in practical terms, a PFAS not only controls the fall to a specified distance but also limits the amount of force an employee is subjected to in the event of a fall. A PFAS consists of an anchorage, connectors, and a body belt or body harness and may include a lanyard, deceleration device, lifeline, or suitable combinations of these. Selecting adequate anchorage points on-site that can withstand shock loads when someone falls is extremely important to the successful use of a PFAS. It is also crucial that body harnesses are fit properly over employees to avoid the creation of slacks that might defeat the purpose of fall protection.

Guardrails consist of top rails, midrails, and toe boards, preventing fall hazards as well as tools or materials falling off scaffolds. OSHA has very specific requirements for the height of the top rails and how much foreign impact the top rail must be able to withstand. Generally, according to OSHA standard 29 CFR 1926.502(b), the top edge height of top rails should be between 39 and 45 inches above the walking/working level. For supported scaffolds in particular, based on OSHA standard 29 CFR 1926.451(g)(4)(ii), this height should be installed between 38 inches (0.97 m) and 45 inches (1.2 m) above the platform surface when the supported scaffolds are manufactured or placed in service after January 1, 2000. In addition, guardrail systems for supported scaffolds should be capable of withstanding, without failure, a force of at least 200 pounds. Although cross bracing such as those shown in Figure 7.10 may serve as a top rail or midrail, there needs to be adequate clearance between the work platform and the cross bracing. Essentially, additional horizontal members are simply added to act as the top rail and midrail.

As previously discussed, the use of head protection will reduce the injury of struck-by hazards. It is sometimes also very helpful to attach yellow warning tapes to specifically delineate the scaffolding area and raise workers' awareness about the potential struck-by hazards.

On FS39, overhead power lines were not present within the site and the hazard of electrocution when erecting and moving scaffolds was minimized. However, it is still an area of concern when projects are in an urban setting or at crowded locations.

Even though general contractors typically subcontract out the masonry proportion of their projects, as the primary site-controlling parties, they ought to know what hazards are associated with masonry construction. Because the FS39 project was under the jurisdiction of the Washington State's safety plan and the potential fall hazard during masonry construction was more than 10 feet, the general contractor should require the masonry subcontractor to furnish a project-specific fall protection plan. Understanding the typical hazards expected during masonry construction will help the general contractor evaluate the masonry subcontactor's fall protection plan.

7.3 STEEL FRAMING

Besides masonry construction, the design firm had selected structural steel construction as the main method for building the two-story station house and the entire fire station roof on FS39. Figures 7.11 and 7.12 are images produced by the building information model introduced in Chapter 5. These images clearly point out how fall hazard continuously posed one of the most significant risks

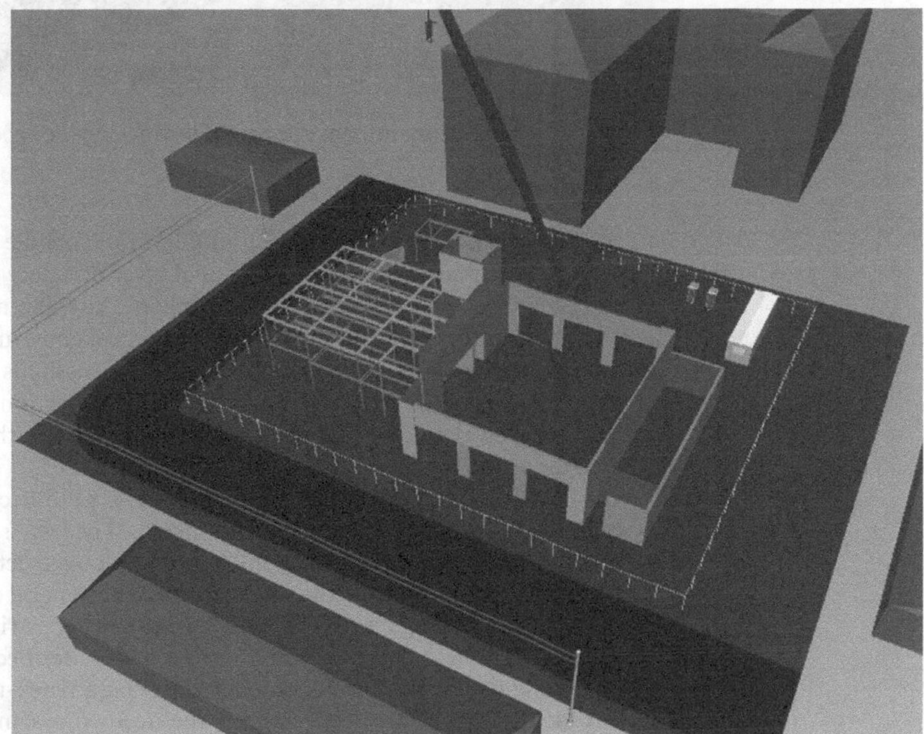

Figure 7.11 Building information model showing the steel work over the two-story station house

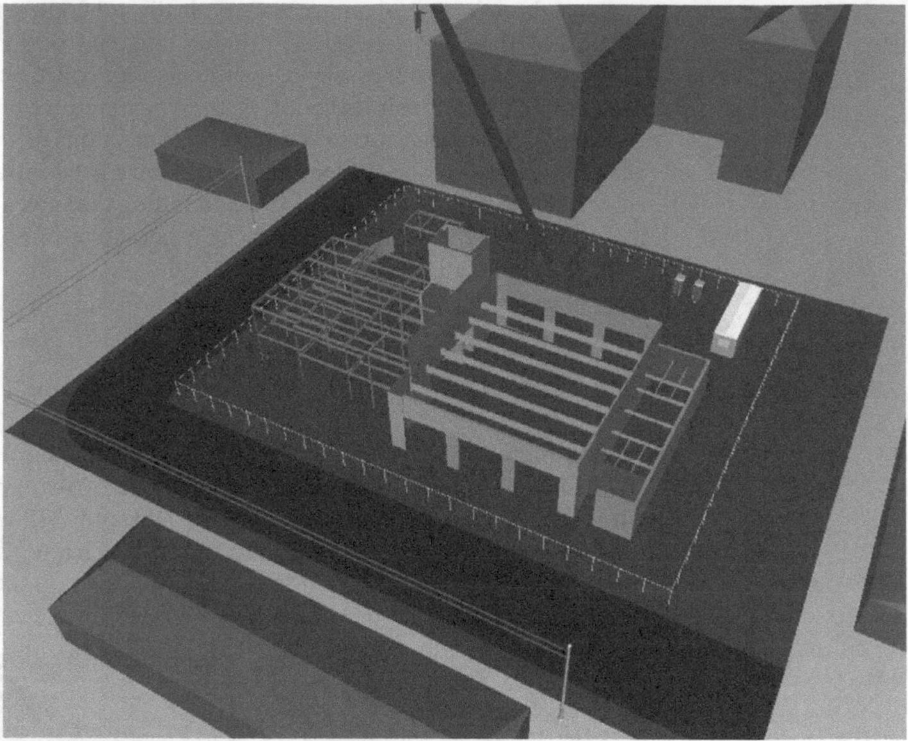

Figure 7.12 Building information model showing the second and third levels of CMU

to the project and the need to have a mobile crane on-site to hoist and place the structural steel, thus creating an additional layer of safety concerns due to workers' exposure to the crane and steel members. Just like masonry construction, general contractors typically subcontract out steel construction to steel erectors. Although steel erectors determine the means and methods of their work, general contractors are still responsible for reviewing the erection and safety plans prepared by the steel erectors and need to be able to assess the hazards involved as well as the custody of safety in relation to each hazard.

On FS39, before steel erection could start, a mobile crane had to be transported to the project site and be mobilized. The general contractor was responsible for providing access points into the construction site (through the northwest site entrance) for the steel erector. The general contractor was also in charge of maintaining a safe control over nearby pedestrian and vehicular traffic when the crane or steel materials entered the site. Because of the relatively limited scale of the project, it is most likely that a truck-mounted, telescoping, boom mobile crane was used without the need to assemble a lattice boom or tower crane. However, safe operations of the crane still required a sufficient clearance with the adjacent structures and an adequate (i.e., firm, graded, and drained) area for the crane. OSHA standards 29 CFR 1926.1407 to 1926.1411 stipulate the clearance that a crane must stay away from the nearby power lines. According to the OSHA standards, there are three options when cranes are operated near power lines. The first option is to deenergize and ground the work site, the second option is to maintain a 20-foot approach distance (or 50 feet when the power line voltage is more than 350 kV), and the third option is to obtain the power line voltage and ensure that the crane and its operations do not get closer than the distances listed in Table 7.2.

Previously, Chapter 5 discussed the overhead power line distribution at the FS39 project site and the crane location in the site logistic plan. It is apparent that the power lines were on the opposite side of the project site and that the mobile crane should be able to maintain at least a 20-foot clearance even when it extended to reach the farthest corners of the site during steel erection. This again reinforces the importance of preproject planning and how a well-sought-out logistics plan can make a difference, especially for steelwork in which the number of crane locations is best kept to a minimum and the laydown areas are apt to be as close as possible to the structure. It is worth noting that OSHA updated its crane standards and made the rules official in November 2010. Among the many updates, power line safety is one of the main changes. The new crane rules also particularly regulate the qualification and certification of crane operators. An operator's certification is valid for five years and can be portable if it is issued by an accredited testing organization.

The area where a crane will operate should have enough bearing capacity to withstand the weight of the crane. Information found in the geotechnical engineering report might be useful for this purpose and the use of blockings under

Table 7.2 Minimum Clearance Distance

Voltage (nominal, kV, alternating current)	**Minimum Clearance Distance (feet)**
Up to 50	10
Over 50 to 200	15
Over 200 to 350	20
Over 350 to 500	25
Over 500 to 750	35
Over 750 to 1000	45
Over 1000	As established by the utility owner/operator or registered professional engineer who is a qualified person with respect to electrical power transmission and distribution

Note: The value that follows "to" is up to and includes that value. For example, over 50 to 200 means up to and including 200 kV.

outriggers would help distribute a crane's load to maintain the crane in a level and stable condition. As the control party on-site, a general contractor must inform the crane operator of known underground conditions (e.g., utilities). In the OSHA new crane rules, information about ground conditions includes all information known about the ground conditions, including written information in possession of the controlling employer (e.g., geotechnical engineering report), whether on-site or off-site. The crane operating area should be barricaded to specifically mark out the crane's swing radius to avoid the struck-by hazard or at the minimum to make other on-site personnel aware of the crane operations.

After a crane is set up and materials delivered, the steel erection can start, but it will also introduce additional hazards such as those associated with working under loads; hoisting, landing, and placing decking; column stability; double connections; landing and placing steel joists; falling objects; and falls to lower levels. According to OSHA standard 1926.753(d), people, other than the employees engaged in the initial connection of the steel or employees necessary for the hooking or unhooking of the load, are not to be under a hoisted load. The standard practically limits the work of other trades during steel erection for safety reasons, and a general contractor's overall project schedule should take into consideration this limitation when sequencing activities that might interfere with steelwork. A realistic example of such a scenario on the FS39 project is the construction of SOG, a non-critical-path activity that could be postponed to after the completion of the steel erection. Figure 7.13 shows the steel work at FS39 and how the reinforcement work of SOG was purposefully scheduled after the steel erection.

During hoisting, lifting multiple steel members (i.e., Christmas tree rigging) of the same shape (e.g., beams) is allowed, given that there is an adequate distance between the steel members and no more than five members are hoisted per lift. The crane operator must observe the equipment's rated capacity and not overload the crane when hoisting materials. If a lift exceeds 75% of the rated capacity of the crane, it is considered as a critical lift and extra attention is required for the safe operation of the lift. A minimum of four anchor bolts per column are required and the general contractor must inform the steel erector that proper curing has been administered and the concrete has reached an adequate strength for the steel column placement. If there has been any repair, replacement, or modification of the anchor rods of a column, the general contractor must provide written notification to the steel erector before the erection of the column. Figure 7.14 shows how a first-floor steel column was secured by four anchor bolts on FS39.

Temporary supports, shoring, guys, bracing, etc., might be necessary to maintain the stability of a steel structure during construction. OSHA's fall protection requirements for steel construction are uniquely situated within its standard 1926.760—except the connectors and employees working in controlled decking zones, each employee engaged in a steel erection activity more than 15 feet above a lower level shall be protected from fall hazards by guardrail systems, safety net systems, personal fall arrest systems, positioning device systems, or fall restraint systems. Deckers between 15 and 30 feet can use a controlled decking zone (CDZ) instead of fall protection. A CDZ is an area in which certain work

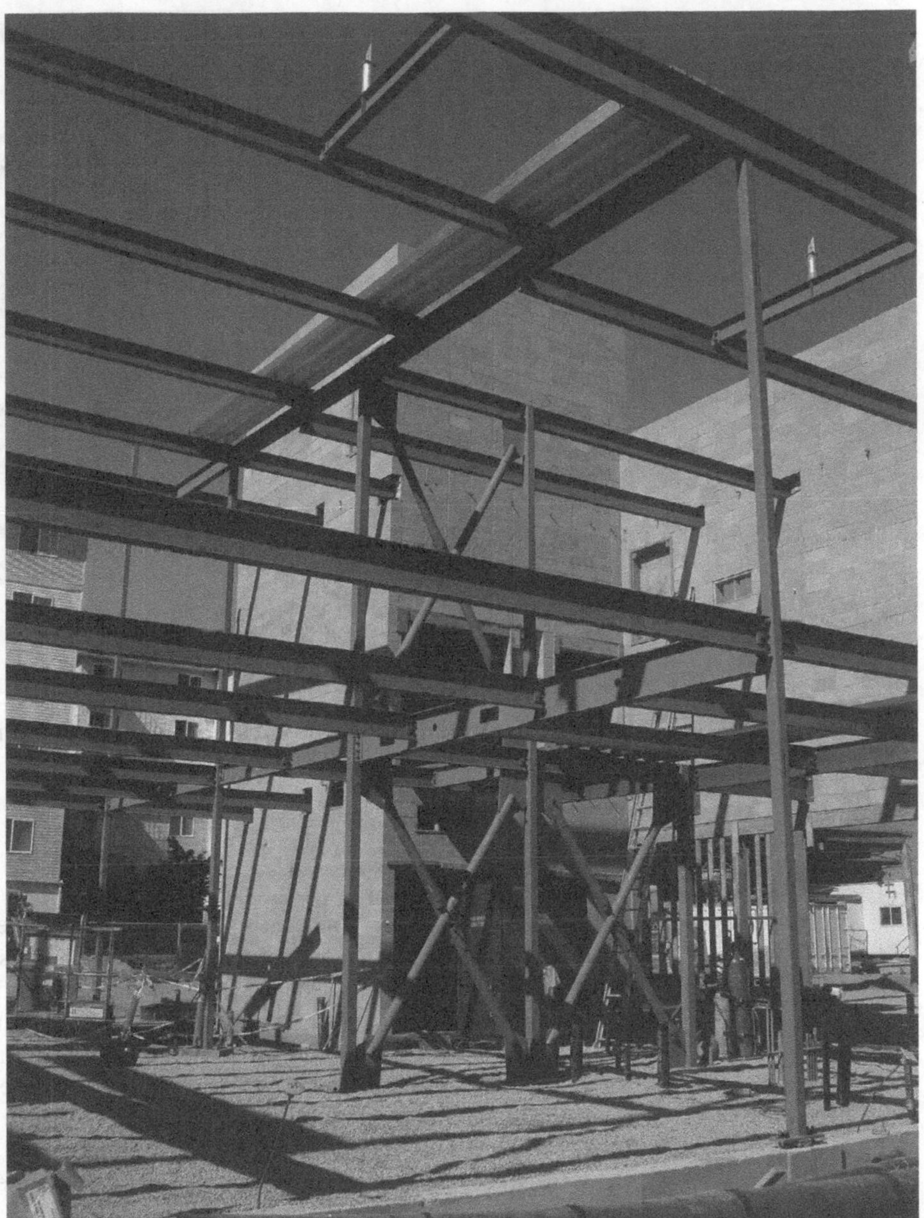

Figure 7.13 Erection of steel at the FS39 two-story station house

(Courtesy of The Miller Hull Partnership, LLP.)

(e.g., initial installation and placement of metal decking) may take place without the use of guardrail systems, personal fall arrest systems, fall restraint systems, or safety net systems and where access to the zone is controlled. However, all iron workers must be protected when the fall exposure is at or more than 30 feet.

Figure 7.14 A minimum of four anchor bolts per column are required per OSHA standards for steel construction

(Courtesy of The Miller Hull Partnership, LLP.)

An additional area of concern is the placement of metal decking, especially considering that the shear connectors often are also the source of tripping hazards. Figure 7.15 shows the shear studs on the second-floor metal decking on FS39 and how the floor perimeter should be protected to prevent falls. These requirements on fall protection are generally the responsibilities of steel erectors. However, a general contractor can direct the steel erector to leave the fall protection in place to be used by other trades. In that case, the general contractor chooses to accept the responsibility for maintaining fall protection equipment left by the erector and should inspect the fall protection equipment prior to authorizing other trades to work in the area.

Miscellaneous equipment or tools used during steel construction include generators, air compressors, welding equipment, and hand tools. Precautions related to the maneuver of such equipment or tools are a part of the everyday safety practices in the field. After the erection of steelwork is completed, the crane has to be demobilized and transported. The same level of attention for the crane to enter into a construction site is now also applicable at this point for the crane to exit the construction site.

Figure 7.15 Tripping hazards caused by shear studs on the metal decking

(Courtesy of The Miller Hull Partnership, LLP.)

7.4 FLOOR CONSTRUCTION

Floor construction may involve slabs that sit on compacted ground (i.e., SOG) or elevated slabs that sit on existing ground. Under a SOG typically resides vapor retarder and subbase granular material to avoid condensation and improve drainage. A contractor needs to identify the locations of plumbing, electrical, and gas services and make rough-in connections before placing the subbase and vapor retarder on-site. Depending on the structural design, a contractor sometimes also needs to build blockouts for the columns before floor construction. After laying down the vapor retarder, a contractor can put in formwork around the slab perimeter and concrete joints and install reinforcement as specified in the project drawing.

Figure 7.16 shows the construction of SOG at the FS39's apparatus bay area after the vapor retarder was placed on-site. Because the steelwork was completed before floor construction, as Figure 7.16 indicates, the roof decks also nicely shield the SOG placement against potential rain. Figure 7.17 illustrates the construction of SOG at the FS39's station building area after the reinforcement was installed. Due to the constant bending and kneeing when tying rebar, employees might experience strains and develop lower back issues over time. As Figure 7.17

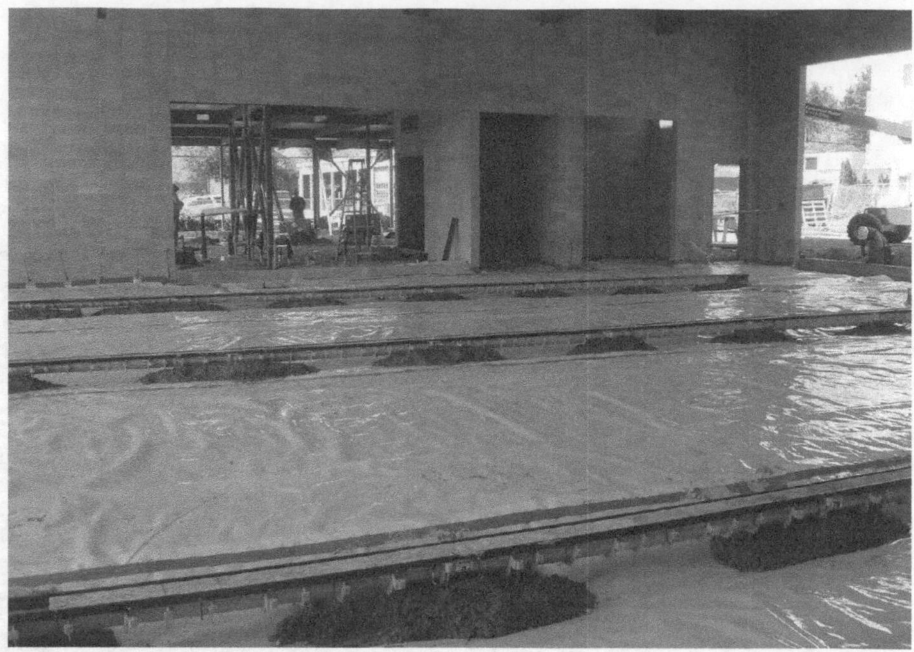

Figure 7.16 Construction of SOG at the apparatus bay area on FS39

(Courtesy of The Miller Hull Partnership, LLP.)

Figure 7.17 Construction of SOG at the two-story station house on FS39

(Courtesy of The Miller Hull Partnership, LLP.)

suggests, it is also very easy for employees to trip and fall because of the rebar, embeds, and blockouts on the ground. Employees who handle the concrete pumping hoses are subject to additional hazards such as getting into contact with the concrete material or being impacted by the sudden high pressure released by kinking concrete hoses.

7.5 SUMMARY

This chapter overviews the typical hazards such as "fall from height," "struck-by," and "electrocution" during superstructure construction. Specifically, the chapter examines in detail the dynamics of masonry construction and steel erection and the hazards they introduce. The use of scaffolding and hoisting equipment during the construction of masonry or steelwork poses additional safety challenges. The chapter also discusses the safety implications to schedule floor construction after the completion of a superstructure as well as the tripping hazards involved during floor construction. Snapshots of the building information model introduced in Chapter 5 are presented in this chapter to highlight the site-specific conditions on the FS39 project and how the conditions translate into criteria that help make meaningful safety planning decisions for superstructure construction.

7.6 REVIEW QUESTIONS

1. How can the building information model be used to determine potential hazards during masonry construction?
2. How can the building information model be used to determine potential hazards during steel erection?
3. What is meant by OSHA Focus Four?
4. How does masonry construction introduce some of the OSHA Focus Four hazards?
5. How can one set up a limited-access zone during masonry construction?
6. How does steel erection introduce some of the OSHA Focus Four hazards?
7. What are the typical components in a guardrail system? What mechanism does the system provide to achieve fall protection?
8. What are the typical components in a personal fall arrest system? What mechanism does the system provide to achieve fall protection?

7.7 EXERCISES

1. Fall protection: Look up OSHA standards 1926.500, Subpart M, "Fall Protection," and list the types of construction activities that are exclusively covered in the other subparts of the OSHA standards. Why are these construction activities listed in the other subparts?

2. Besides the fabricated tubular welded frame scaffold, what other types of scaffolds are common in construction and what are their related safety precautions?
3. Leading edge refers to the unprotected side and edge of a floor, roof, or formwork for a floor or other walking/working surface (such as deck) which changes location as additional floor, roof, decking, or formwork sections are placed, formed, or constructed. Where would the leading-edge work take place in the FS39 project? What OSHA standards are applicable to leading-edge work? Based on the OSHA standards, who (the steel erector or the controlling general contractor) has the custody of fall protection?
4. Under what situation is the use of guardrails preferred over the use of a PFAS and why? Under what situations is the use of a PFAS preferred over the use of guardrails and why?
5. Other than guardrails and PFASs, what other acceptable means of fall protection are available?
6. According to OSHA, other than exceeding the 75% rated capacity of a crane, what other types of crane operations meet the definition of a critical lift?

8

SAFETY FOR THE EXTERIOR ENCLOSURE

8.1 INTRODUCTION

Once the basic structure of the building was completed, the next set of construction tasks related to the exterior enclosure. This involved the installation of wood framing and sheathing on two floors, installation of waterproof membrane, installation of metal and cement fiber siding, construction of the roof, and installation of the glazing. This chapter addresses some of the safety risks faced by construction workers during this phase of the project and measures that should be taken to minimize the potential for injury.

8.2 WOOD STUD FRAMING AND SHEATHING

Figure 8.1 shows the location of the wood framing on the first floor, and Figure 8.2 shows the location of the wood framing on the second floor. As previously discussed, the building information model can be used to plan the execution of this work safely. Wood frame sections were constructed on the second floor, as shown in Figure 8.3, and were then raised into position around the perimeter of the building. This allowed the framing construction to occur on a level plane not requiring the use of ladders. Construction of the second floor was nearing completion in Figure 8.4. After the second-floor framing was complete, the contractor moved to the first floor to install the framing in a similar manner.

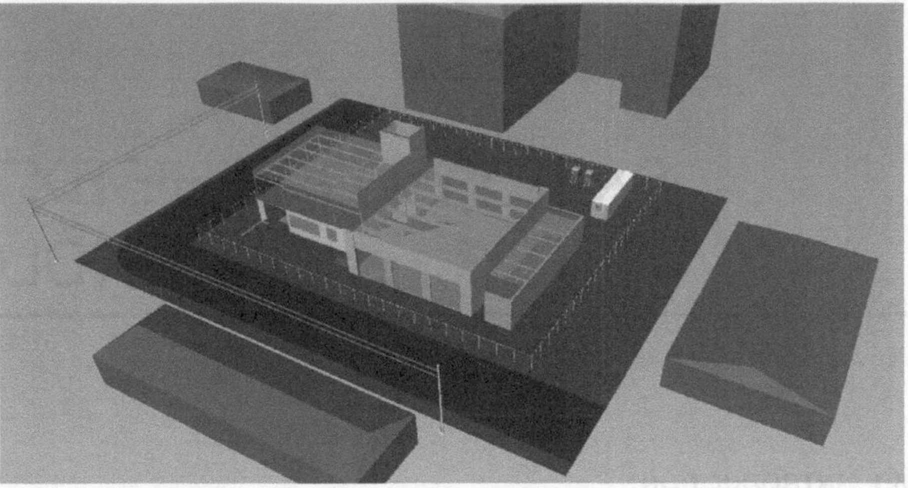

Figure 8.1 Building information model showing wood framing on first floor

To minimize the potential for injury, all workers need to wear PPE (safety glasses, gloves, and hard hats). Hand and power tools need to be inspected for proper maintenance, and workers need to be trained on their proper use. All working surfaces need to be kept clean and dry to prevent accidental tripping or slipping. All power tools need to be grounded and disconnected when not in use.

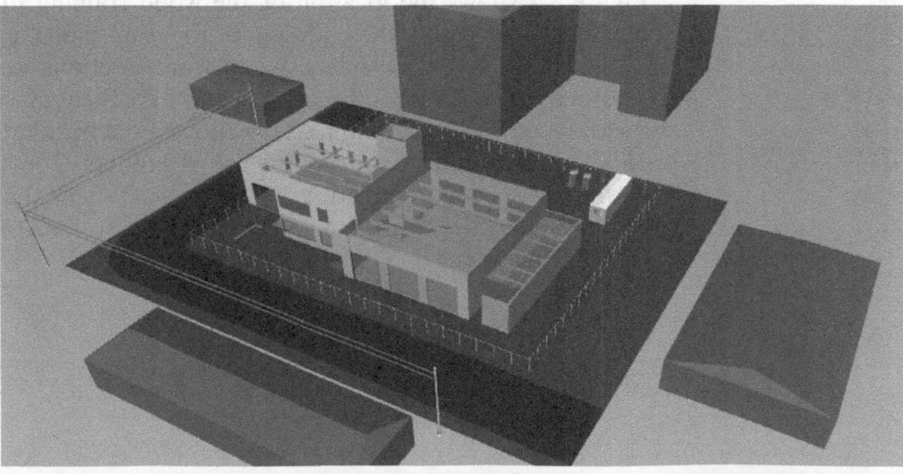

Figure 8.2 Building information model showing wood framing on second floor

Figure 8.3 Second-floor wood frame section construction

(Courtesy of The Miller Hull Partnership, LLP.)

Figure 8.4 Second-floor wood frame

(Courtesy of The Miller Hull Partnership, LLP.)

Figure 8.5 Wood sheathing on first and second floors

(Courtesy of The Miller Hull Partnership, LLP.)

Figure 8.6 Temporary guardrails on the roof perimeter

(Courtesy of The Miller Hull Partnership, LLP.)

Once the framing had been completed, construction workers installed the wood sheathing using a powered man lift as shown in Figure 8.5. Proper fall protection is required to minimize the potential for falling, and no one should be allowed to work under the man lift to preclude being struck by falling objects. To prevent the workers who were installing wood sheathing around the hose tower on the rooftop, temporary guardrails were erected at the roof perimeter as shown in Figure 8.6.

8.3 SIDING INSTALLATION

After the sheathing was installed, waterproofing was placed on the sheathing and the siding installed from the man lift as shown in Figure 8.7. As addressed in the APP shown in Appendix B, workers in the man lift must stand firmly in the floor of the basket and wear a body belt or harness that is fastened to the basket with a lanyard. This is illustrated in Figure 8.7. The man lift cannot be moved when the boom is elevated in a working position with workers in the basket. The brakes must be set whenever the man lift is in use.

Figure 8.7 Metal siding installation

(Courtesy of The Miller Hull Partnership, LLP.)

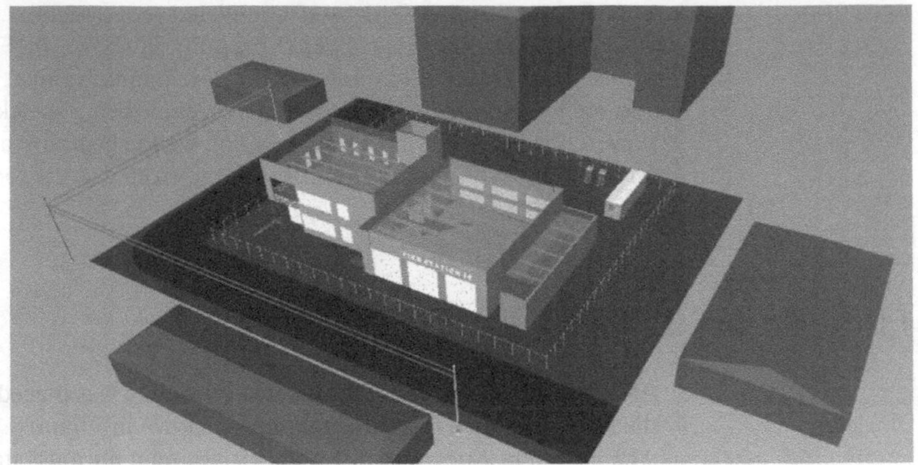

Figure 8.8 Building information model showing glazing

Figure 8.9 Metal frames to support glazing installation

(Courtesy of The Miller Hull Partnership, LLP.)

8.4 GLAZING INSTALLATION

The building information model shown in Figure 8.8 can be used to plan the installation of the glazing. Metal frames were installed in the openings, as shown in Figure 8.9, and the glazing was installed within the frame. Care must be exercised to ensure that the glass sections do not fall and break creating a safety hazard. Workers handling the glass must wear gloves to protect their hands.

8.5 ROOF CONSTRUCTION

A built-up roof was used on the fire station. The locations of the roof sections are shown in the building information model in Figure 8.10. The first step was to install the metal decking shown in Figure 8.11. The metal deck sections were placed on the steel structure with the crane and by construction workers. To minimize the risk of falling workers were required to wear body harnesses and be secured to the structure with lanyards. This was followed by placement of the insulation shown in Figure 8.12. Once the insulation was installed, the waterproof membrane was installed to complete the roof as shown in Figures 8.13 and 8.14. One end of the roof construction site was totally surrounded by walls and posed little danger of workers or materials falling. However, as Figure 8.14 suggests, some area of the roof parapet wall was still quite low and could cause fall hazards. In addition, as illustrated in Figure 8.15, the skylights on the rooftop were not guarded, introducing additional hazards because employees could easily fall through the skylights and be severely injured.

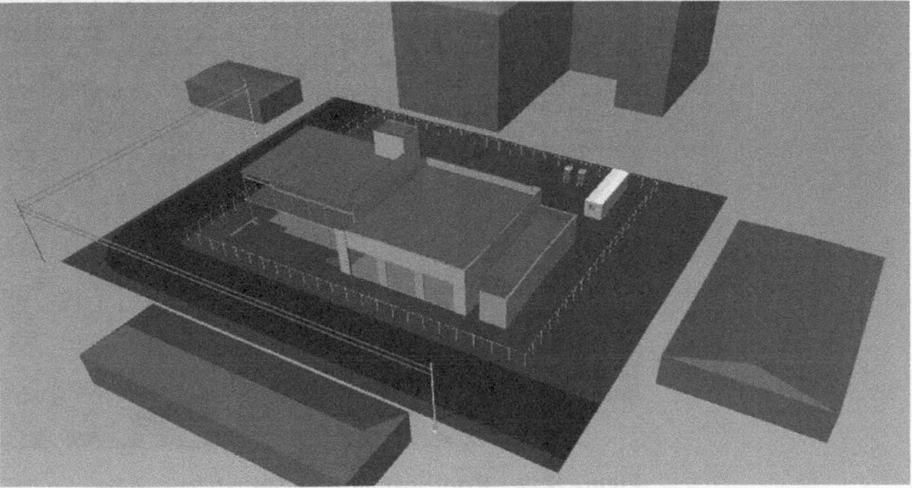

Figure 8.10 Building information model showing roof locations

Figure 8.11 Metal roof deck

(Courtesy of The Miller Hull Partnership, LLP.)

Figure 8.12 Roof insulation installation

(Courtesy of The Miller Hull Partnership, LLP.)

Figure 8.13 Waterproof membrane installation

(Courtesy of The Miller Hull Partnership, LLP.)

Figure 8.14 Completed roof

(Courtesy of The Miller Hull Partnership, LLP.)

Figure 8.15 Skylights on the roof

(Courtesy of The Miller Hull Partnership, LLP.)

8.6 SUMMARY

The major safety hazards associated with the construction of the exterior enclosure related to the installation of the wood framing, sheathing, waterproofing, and siding. The wood framing was constructed on a flat surface to avoid the hazard posed by the use of ladders. The installation of the sheathing, waterproofing, and siding was performed from a man lift using proper safety procedures including proper PPE. Roof construction on this project did not present the typical falling hazard because the work area was enclosed within walls. However, special care was needed during the placement of the metal roof decking to ensure that workers did not fall while it was being installed.

8.7 REVIEW QUESTIONS

1. What are the major safety hazards associated with wood frame construction?
2. What actions can be taken to mitigate the hazards identified in Question 1?

3. What is a man lift, and what are the major safety hazards associated with its use on a project?
4. What are the major safety hazards associated with glazing construction?
5. What are the major safety hazards associated with roof construction?

8.8 EXERCISES

1. Research the types of body harnesses and lanyards that are available for wear by people working in a man lift basket?
2. Develop a safety procedure to be followed if wood frame walls must be constructed vertically.
3. Research the proper safety procedures that must be followed if workers are constructing a steep-pitch roof.

9

SAFETY FOR THE INTERIOR CONSTRUCTION

9.1 INTRODUCTION

Logistically, while a building's exterior enclosure is taking place, its interior is well protected from the weather and specialty contractors can start the installation of MEP (mechanical, clectrical, and plumbing) rough-ins. After major MEP rough-ins are completed, wall and ceiling framing are assembled before drywalls are installed, primed, and painted. Internal doors, hardware, lavatory, and fixtures are then installed subsequently. Follow-ups are the flooring finish and interior cleanup before the final punch. This project phase is typically referred to as interior construction and is a stage in which different trades interfere with each other and compete for the available space on-site. Meticulous sequencing and activity coordination are crucial to the successful completion of interior construction.

Because a vast amount of horizontal pipelines, ducts, and conduits are located above the ceiling, overhead struck-by hazards are common. In order to reach sufficient height for the installation of MEP rough-ins and other overhead work, ladders, mobile scaffolds, and scissor lifts are often employed on-site during interior construction. This introduces fall hazards at locations that are not normally considered dangerous by the workers. For installing vertical pipelines, ducts, and wires within shafts, the concern for fall hazards is even higher. Although exterior enclosure such as roofing and sheathing provides protection for the interior construction activities, it could also reduce the amount of indoor lighting before the light fixture is operable or create enclosed areas where ventilation can be challenging for plasters and painters. Various power tools are continuously being used with cords and wires presenting tripping hazards on-site. Materials such as drywall stored inside could take up space that is already limited and make the site more congested. All in all, housekeeping becomes more critical and is often a leading indicator of a site's overall safety performance.

Specifically on the FS39 project, interior construction activities included the rough-ins of MEP systems, floor sanding, wall and ceiling framing, doorjamb setting, the rough-ins of walls and ceilings, drywall installation, paneling for walls and ceilings, interior painting, floor covering, door and hardware installation, casework installation, lavatory and fixture installation, wheelchair lift installation, and final finish.

9.2 MEP ROUGH-IN

MEP is an acronym that traditionally represents the mechanical, electrical, and plumbing systems in a construction project. However, with the ever-expanding scope and complexity of nowadays construction projects, the scope of MEP could also include fire protection, security/safety control, process piping, telecommunication, and other types of systems. MEP rough-in is the process in which specialty contractors bring in major MEP systems and lay out the locations of basic lines and components without making the final connections. For HVAC (heating, ventilation, and air conditioning) systems, this could mean placing an air conditioning unit on the rooftop and installing the duct work, piping, tubing, fitting, valves, terminal boxes, and other distribution components. For electrical systems, rough-in could include the installation of service mains, wiring, panels, conduits, fittings, outlet boxes, cables, and transformers and pulling cable through conduit and splicing in electrical boxes. In the case of plumbing systems, rough-in means the installation of the drain, waste, and water supply lines based on the proposed location of each fixture. The process of rough-ins helps specialty contractors coordinate the physical requirements of their systems and detect potential clashes earlier on. More and more projects are using BIM nowadays for MEP coordination as expensive mistakes can be detected on screen and be avoided on-site.

Figure 9.1 shows two condensation units installed on FS39's rooftop at the station house during MEP rough-in. Even though the rooftop was sloped toward the south end of the fire station, the parapet wall was still 5 to 7 feet above the roof and could serve as a means of fall protection. This also exemplifies what the design community could do to promote construction safety subtly. Walking pads around the condensation units were installed to protect the bituminous membrane roofing with the expectation that future maintenance work might introduce heavy frictions to the roof.

Figure 9.2 illustrates the rough-in installation of the HVAC system at the FS39's apparatus bay area. As Figure 9.2 displays, during MEP rough-ins the project site can be challenged by the overhead hazards, fall hazards (introduced by the use of ladders and scissor lifts), site congestion, and housekeeping issues. Overhead hazards are for both field personnel and equipment, as personnel could be injured and equipment could damage the installed systems when in contact with installed MEP rough-ins. Figure 9.3 shows how red warning tapes were used as means of precautions for the low-hanging pipelines in a parking garage project that the authors visited.

Figure 9.1 Condensation units installed during FS39 rough-in

(Courtesy of The Miller Hull Partnership, LLP.)

Figure 9.2 Rough-in installation of the HVAC system on the FS39 project

(Courtesy of The Miller Hull Partnership, LLP.)

Figure 9.3 Using warning tapes to specifically mark out the overhead hazard

(Photo taken by the authors.)

Portable ladders are one of the most commonly used tools on construction job sites, especially during interior construction. However, according to the 1997 "Death from Falls in Construction" report developed by the Center for Construction Research and Training (CPWR), ladders are also one of the top three locations where workers fell to their deaths. Although the report from CPWR was dated back to 1997, the fatality pattern has not changed much and ladders continue to be on the center of attention. In order to reduce the potential hazards associated with ladder uses, some contractors even opt to avoid ladders and turn to other means of accessing higher levels as long as the on-site situations warrant such decisions. OSHA standard 29 CFR 1926 Subpart X specifies the safe working requirements that apply to all ladders, including both manufactured and job-made ladders.

Choosing the right ladder for the job task at hand is the number one step to prevent ladder misuses. Non-self-supporting ladders such as straight ladders and extension ladders are adequate for accessing a higher work surface but self-supporting ladders such as A-frame ladders (a.k.a. step ladders) are more appropriate when there is no solid structure for the ladders to lean against. The ladder to the left in Figure 9.4 is an example of a straight ladder whereas the ladders in the middle and to the right in Figure 9.4 are examples of step ladders. Capacity-wise, ladders come in different duty ratings and various heights. Only those which are capable of handling the job task should be used. Common issues related to

Figure 9.4 Ladder uses during interior construction on FS39

(Courtesy of The Miller Hull Partnership, LLP.)

ladder uses include improper ladder setup, portable ladders not extending 3 feet above landing surfaces, ladders not being secured properly (i.e., tie-offs), employees standing on the top two rungs of a step ladder, and employees overreaching when working from a ladder. Informing workers about the ladder hazards and ways to avoid the hazards through training is integral to addressing these common issues. To set up a ladder properly, the ladder needs to first stand on a stable, level, and nonslippery footing. Additionally, a simple rule of thumb for setting up a non-self-supporting ladder is to apply the 4:1 (rise-over- run) ratio which stabilizes the ladder while providing a practical slope for ascending and descending. Workers ascending or descending a ladder must face the ladder and maintain a three-point contact. Besides proper setup, ladder locations also matter. During interior construction, partition walls are up and doors are installed. If during this stage a ladder is placed at high-traffic areas where workers are passing through or in front of doors that can open out into the ladder, it is very likely that someone will unintentionally displace the ladder.

Besides ladders, scissor lifts and mobile scaffolds are also often employed during interior construction to access higher levels of work areas. Scissor lifts are not covered by aerial lift—related provisions in OSHA standard 29 CFR 1926, Subpart L, "Scaffolds," because, per the ANSI definition, they are not considered aerial lifts. However, from the regulation's point of view, scissor lifts are still a type of scaffold and should comply with applicable provisions from Subpart L. In addition, since scissor lifts are mobile, the specific requirements from OSHA standard 29 CFR 1926.452(w) for mobile scaffolds should specifically be observed. Precautions for working on scissor lifts or mobile scaffolds are that casters or wheels should be locked and that employees are not allowed to ride on scaffolds unless certain conditions exist. Having adequate means to reach the scaffold platform is always a must. Figure 9.5 shows how an employee was using an inadequate access to get onto a mobile scaffold at a commercial headquarter project that the authors visited.

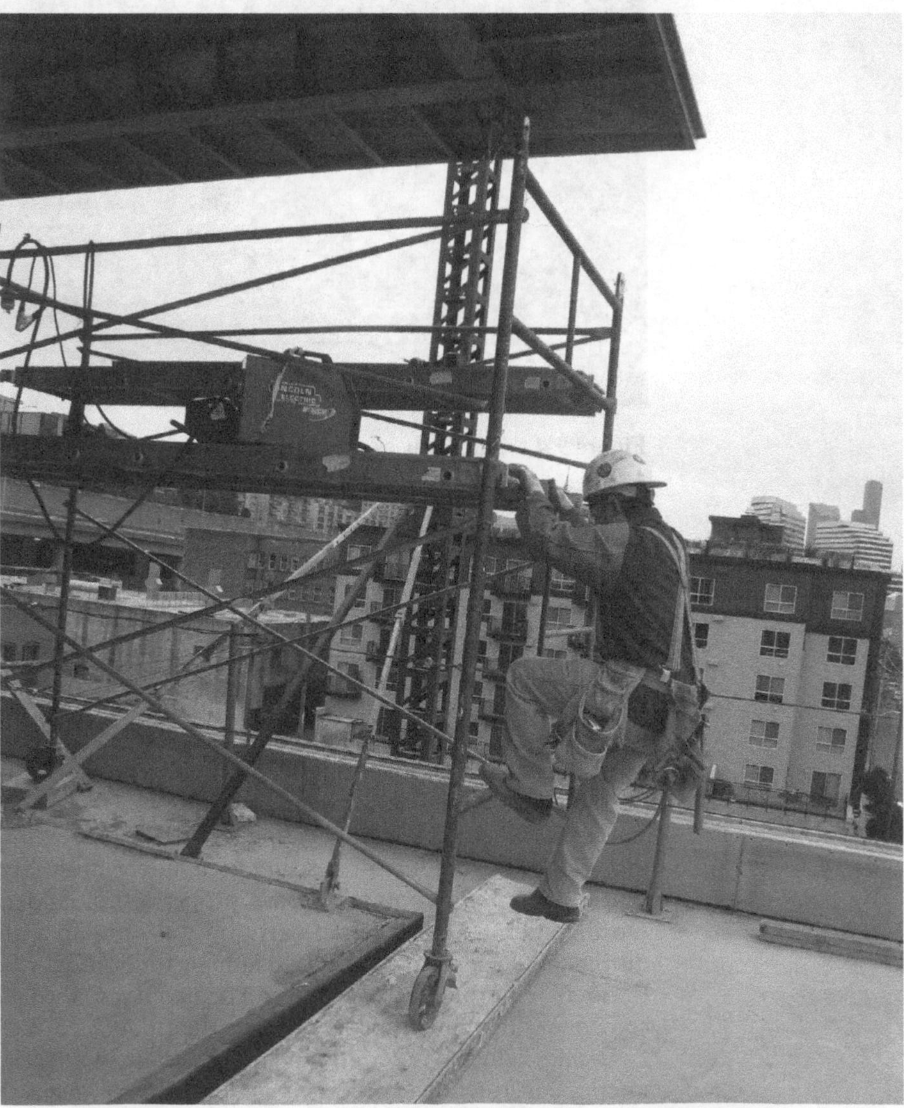

Figure 9.5 Inadequate access to a mobile scaffold

(Photo taken by the authors.)

Because FS39 was a new construction project, concerns over existing MEP systems were nonexistent. However, for renovation type of projects, extra care must be given when connecting to or upgrading existing systems as electrical hazards could cause severe injuries if a lock-out-tag-out (LOTO) procedure is not established or followed. OSHA standards related to the lock-out and tagging of circuits for construction are in 29 CFR 1926 Subpart K.

Figure 9.6 illustrates FS39's site congestion and housekeeping condition with a scissor lift parked in the middle of the figure. Housekeeping is often a leading indicator of a site's overall safety performance and an aspect that safety managers always inspect when they perform job walks. OSHA's housekeeping-related regulations are in 29 CFR 1926 Subpart C and mainly concern the collection and removal of debris, scrap, and waste. For example, spills should be cleaned up as soon as possible, especially if they are slippery or hazardous liquids. In practice, it is equally important to keep stairways and walkways clear of materials, cords, and wires to reduce the tripping hazards. Raising cords and wires above ground also minimizes their tear and wear against the sharp edges of light-gauge metal framing typically used for interior partitioning. Loose materials should be secured from unexpected falling or breakage.

Figure 9.6 FS39's site congestion and housekeeping condition during interior construction

(Courtesy of The Miller Hull Partnership, LLP.)

It is apparent that even though specialty contractors are responsible for the safe working practices within their scopes of work during MEP rough-ins, there are many shared, common hazards that a general contractor should watch out for and exercise its control over.

9.3 WALL AND CEILING FRAMING

At FS39's station house, the framing of interior walls was mostly completed with wood studs except where the metal studs presented near shaft walls. The ceiling system was a combination of suspended acoustical ceiling tiles (ACTs), wood panels, and gypsum wall boards (GWBs). Figure 9.7 shows where the interior framing and MEP rough-ins had to be coordinated. Electrical rough-ins usually take place after framing has been completed and before drywalls have gone up. During wall and ceiling framing, all hand and power tools need to be inspected for proper maintenance, and employees need to be trained on their proper use. Power tools such as nail guns and circular saws are heavily used, which makes proper PPE such as eye protection, steel-toe boots, hard hats, and hearing protection particularly important for both the tool operating workers as well as the nearby workers. In the event of unintentional nail gun firing or nails penetrating through the target surface, nearby workers could be severely injured. For eye protection, regular prescription lenses cannot resist high-impact

Figure 9.7 FS39's interior wall framing

(Courtesy of The Miller Hull Partnership, LLP.)

Figure 9.8 FS39's installed drywalls
(Courtesy of The Miller Hull Partnership, LLP.)

forces and do not constitute appropriate eye protection. Acceptable safety glasses or goggles must meet the ANSI standards Z87.1 if they were purchased after July 1994. A portable circular saw must have both upper and lower guards and the lower guard must automatically return to the covering position when the saw is withdrawn from the work. Workers should avoid carrying the power tools by their cords and good housekeeping continues to help reduce tripping or slipping hazards caused by the tangling cords on the ground. Similar to MEP rough-ins, ladders and mobile scaffolds are also used to raise the working height, and previously discussed precautions should be observed to avoid fall hazards. Once the framing has been completed, construction workers can install the drywalls for taping and painting. Figure 9.8 shows an enclosed area on FS39 where the drywalls had been installed.

9.4 PAINTING AND FLOOR COVERING

In Figure 9.8, employees were using a ladder for wall and ceiling painting. It is also quite common to see plasters and painters work on stilts for easy mobility and for accessing higher ceilings or walls. Stilts are a pair of poles or similar supports with raised footrests used to permit walking above the ground or working surface. Similar to ladders, stilts are easy-to-operate tools that increase employee productivity but they too introduce fall hazards along with their uses. Employees are advised not to wire stilts together to bump up their heights. As regulated by OSHA standard 29 CFR 1926.452(y), "Any alteration of the original equipment shall be approved by the manufacturer."

Figure 9.8 also shows how exterior enclosure influences interior construction in terms of an enclosed area's lighting condition and ventilation capacity. The exterior sheathing provided protection for the glazing openings, but it reduced the natural lighting and made task lighting necessary. Because interior wall framing had been completed, the enclosed area had no natural ventilation and could cause health concerns for employees who were handling finishing compound or paint solutions. For example, high exposure to joint compound for drywall finishing could cause both acute and chronic inhalation hazards when the ventilation is poor at a work site. For this reason, the MSDS or SDS of joint compound or paint solutions typically recommends the use of NIOSH-approved respirators if engineering controls cannot be implemented to provide effective ventilation. The nature of the containment and the permissible exposure limit (PEL) determine the types of respirators that are appropriate. Respirator-related regulations can be found in OSHA standards CFR 1910.134. The standards apply to not only the general but also construction and other industries.

After interior surfaces have been painted, the installation of finish flooring materials usually takes place as one of the last few activities in order to protect the flooring materials. On FS39, the interior wall assembly finish schedule in the architecture drawing includes the concrete, carpet, tile, rubber, and grating finishes for the project flooring. Figure 9.9 shows a section of the apparatus bay

Figure 9.9 Concrete flooring

(Courtesy of The Miller Hull Partnership, LLP.)

concrete floor after it was sanded by a grinding machine. Power equipment like grinding machines may come with the so-called dead-man switches to protect the operators from machines getting out of control. When OSHA refers to manufacturer recommendations in the standards, using machines without recommended safety mechanisms like the dead-man switches can be a violation.

9.5 MEP FINISH

After the light fixtures, mechanical units, and plumbing fixtures are set, the final stage of MEP work follows next. Plumbing fixtures such as water closets, urinals, showers, and sinks are set and connected. In parallel, bathroom partitions are installed and final water line connection is made to the building. Electric finishing and trimming involve installing and connecting receptacles, switches, light fixtures, light-duty devices, heavy-duty utility devices, controls, and appliances. Electric final also includes circuit testing to ensure proper connections and the compatibility of installed components. For HVAC, the installed systems are commissioned through testing, adjusting, and balancing. During MEP finish, specialty contractors must adhere to the sequence of testing recommended by the corresponding system manufacturers. Electrical shocks continue to pose some concerns while the project wraps up for the final punch and inspection.

9.6 SUMMARY

Interior construction starts after a building's exterior enclosure can provide protection against weathering for the activities inside the building. It is a stage in which different trades interfere with each other and compete for the available space on-site. On the FS39 project, major activities that took place during interior construction include MEP rough-ins, wall and ceiling framing, interior painting, flooring, and MEP finish. During MEP rough-ins, a project can be challenged by the overhead hazards, fall hazards, site congestion, and housekeeping. The use of portable ladder, scissor lifts, and mobile scaffolds for accessing higher levels of working surface is the main source of fall hazards. Knowing how to choose the right ladder for the job task at hand through proper training is the first step to prevent ladder misuses. Precautions for working on scissor lifts or mobile scaffolds are that casters or wheels should be locked. There should also be adequate means to reach the lift or scaffold platform. In addition, the number of trades involved and the range of materials stacked indoors make project sites very congested during interior construction. Housekeeping becomes critical and all field personnel should be mindful about the timely removal of debris as well as clearing passageways by raising wires and cords. The safe use of power tools such as nail guns and circular saws along with the employment of necessary PPEs are essential during interior wall and ceiling framing. Because exterior enclosure reduces indoor lighting and natural ventilation, workers conducting interior painting should have proper task lighting and respiratory protection, following the recommendations specified in related MSDSs. Another potential concern during interior painting is the fall hazard introduced by the use of stilts. Finally, specialty contractors complete the setting and connection of MEP systems during MEP finishes. Potentially, electrical shocks could still happen during MEP finishes, and specialty contractors should adhere to the testing and commissioning procedure specified by the system manufacturers.

9.7 REVIEW QUESTIONS

1. What are the typical hazards seen during MEP rough-ins and why?
2. What are the common issues associated with the use of portable ladders?
3. How can a self-supporting ladder be set up properly?
4. How can a non-self-supporting ladder be set up properly?
5. What are considered as housekeeping activities?
6. What does a "dead-man switch" mean?

9.8 EXERCISES

1. Based on OSHA standard 29 CFR 1926.452(w), what are the precautions for working on mobile scaffolds?
2. What is a lock-out-tag-out procedure and why is it essential for ensuring the safe operation and testing of electrical equipment?
3. Visit two construction sites in your neighborhood and observe how the housekeeping conditions on the two sites differ. What are the best practices you see and what areas are slipped?
4. What hazards are related to the use of pneumatic tools such as nail guns? What PPEs should be used to protect workers from these hazards?
5. Find the MSDS or SDS of any common paint and discuss how engineering controls and proper PPEs protect exposed workers from related material hazards.

10

SAFETY FOR THE MISCELLANEOUS WORK ITEMS

10.1 INTRODUCTION

As a project rapidly marches through major construction activities and towards its final site work, its general contractor also needs to attend to miscellaneous work items in the last construction phase in order to wrap up the project. On the FS39 project, such miscellaneous work items are illustrated in Figure 10.1 and included the fueling and garbage recycling shelter, landscaping, and installation of a steel artwork. It is worth noting that not all these items were in the general contractor's scope of work. However, as the lead party on-site, the general contractor was still responsible for the overall site safety, especially at places where the miscellaneous work items were integrated into the completed structure. As discussed in FS39's site logistic plan in Chapter 5, the northwest corner of the site was the main entrance for equipment and material hauling and delivery. For this reason, the fueling and garbage recycling shelter was one of the few final work items so that the general contractor could keep the site entrance clear for other activities. A snapshot of the FS39 building information model showing how the shelter was scheduled after major construction activities and before landscaping were completed is shown in Figure 10.2. Landscaping included the north and south garden areas, which called for granite curbing stones at various locations, posting ergonomic challenges for the landscaping workers. The galvanized metal structure artwork was designed to sit on top of the underground cistern structure for receiving the stormwater runoff and transporting it into the cistern. Although the metal structure artwork was fabricated and placed by the project owner's contractor, the general contractor had to coordinate with the owner's contractor and administered traffic control during the artwork installation. Figure 5.1 in Chapter 5 illustrates the 28-ft-tall artwork next to the rain garden.

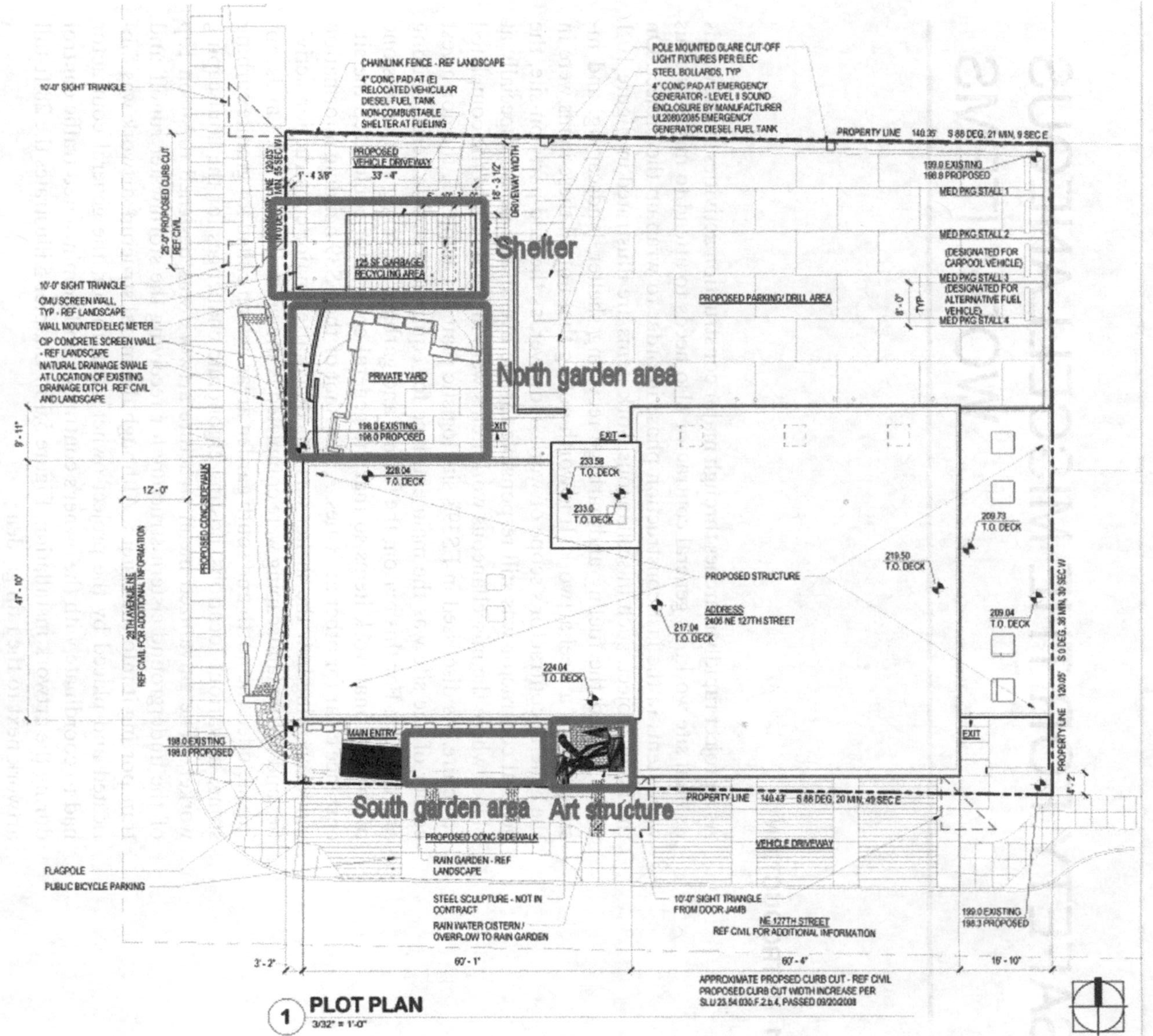

Figure 10.1 FS39's plot plan showing the locations of miscellaneous work items on-site

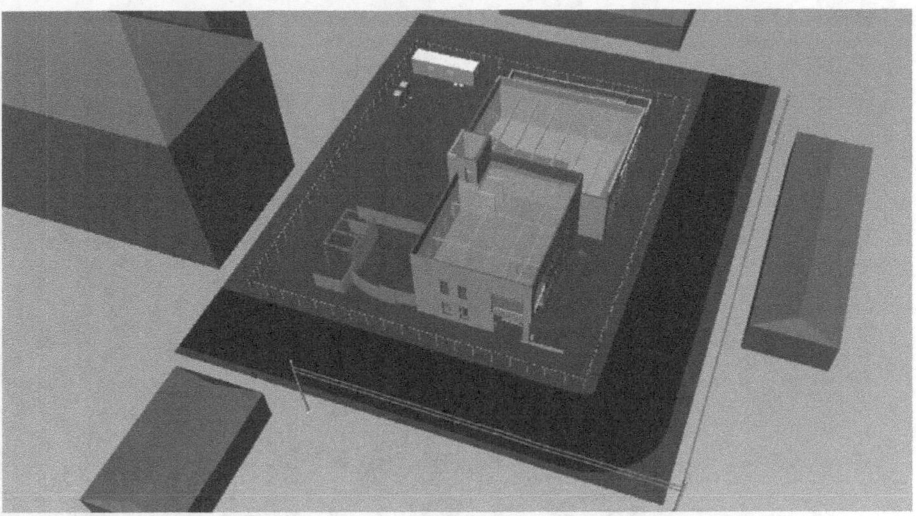

Figure 10.2 Building information model showing the shelter location

10.2 FUELING AND GARBAGE RECYCLING SHELTER

The shelter was composed of a masonry wall, metal siding on the back of the wall, hollow structural steel (HSS) posts, a metal roof, and some structural steel to support the roof. The shallow continuous footing underneath the shelter was reinforced and built with cast-in-place concrete during foundation construction, even though it was a part of the site work. By the time the shelter's masonry wall was assembled, the majority of the FS39 superstructure had been in place. Masonry construction and its related hazards were discussed in detail in Chapter 7 and most of the reviewed hazard mitigation controls are still applicable except that scaffolding was not required. The steelwork proportion of the shelter was installed last before landscaping started. To bolt and connect the steel members on the shelter roof, a worker would most likely use an extension ladder to access the rooftop. Precaution on the proper ladder setup and uses remained an important topic and training should be given to the employee engaging in the roof assembly. Because the top of the shelter was only 13 feet above ground, OSHA's fall protection rule for steel erection did not apply. In 29 CFR 1926.760, OSHA stated that "each employee engaged in a steel erection activity who is on a walking/working surface with an unprotected side or edge more than 15 feet (4.6 m) above a lower level shall be protected from fall hazards by guardrail systems, safety net systems, personal fall arrest systems, positioning device systems or fall restraint systems." However, in reality, fall hazard continued its presence, and since the fire station project fell under the Washington State's plan, an employee standing on the shelter roof should still have a proper means of fall protection.

Figure 10.3 shows how a worker was connecting the channels at the edge of the FS39 shelter roof. An extension ladder was set up on wooden sheathings in

Figure 10.3 A worker connecting the steel channels on the shelter roof at FS39 (Courtesy of The Miller Hull Partnership, LLP.)

order to gain access to the roof. The sheathings could get slippery when being wet and reduce the stability of the ladder. In addition, it would be difficult for the worker to maintain three points of contact when ascending the ladder and carrying the steel channel at the same time if no helper was on-site to provide assistance. For fall protection purposes, considering the elongation distance of a fall arrest system, a fall restraint system would be a better choice in this case. All in all, as a project is close to its final completion, field personnel can easily drop their guard against potential hazards. Persistent communication on the importance of work safety and health will help reinforce employees' awareness about their surroundings and drive the safety culture even after a project is completed.

10.3 LANDSCAPING

There were two gardening areas at FS39, including the north garden, which contained an enclosed yard, and the south garden, which was designed to serve as the rain garden by absorbing the overflow of collected rainwater runoff. The work scope of landscaping included not only planting around the two gardens but also grading, paving, surfacing, back filling, and installing the irrigation system. Figure 10.4 captures the scene in which two landscaping employees were working on

Figure 10.4 North garden area granite stone seating placement

(Courtesy of The Miller Hull Partnership, LLP.)

the granite stone seating in the north garden after the fueling and garbage recycling shelter was completed. Figure 10.5 shows how the curbstone on the north end of the south garden was completed much earlier in order to coordinate with other utility and site work.

Soil preparation, grading, and backfilling introduce hazards such as trips and slips as well as hazards that are related to the use of grading equipment. Paving and surfacing, if involving manual material placement and orientation, can create ergonomic challenges because the workers spend a significant amount of time on their hands and knees. Similar ergonomic challenges apply when it comes to planting activities, and continued stress on hands and knees can introduce bursitis, tendinitis, or arthritis. Additionally, lifting heavy granite stones, planters, or trees increases the chance of musculoskeletal disorders and in the event of a dropped load could cause bruises or fractures. Workers should recognize their limits, be trained in proper lifting techniques, use engineering or administrative controls (when possible), or work as a team. The price for curing a musculoskeletal disorder is quite dear for the injured employee, the employer, and the workers' compensation insurance. The road to recovery could be a long battle and the injury might even follow the employee for life. Some construction sites spend the

Figure 10.5 South garden curbstone placement

(Courtesy of The Miller Hull Partnership, LLP.)

first 5 to 10 minutes of each work day to administer the so-called stretch-and-flex program in order to improve worker flexibility and avoid sprains and strains. For installing the irrigation system, the landscaping subcontractor must be familiar with all utility, civil, and electrical plans so that the required trenching operations do not damage embedded lines or cause electrical hazards.

10.4 ARTWORK

The galvanized metal art sculpture at the FS39 was designed to redirect the rain-water runoff into the underground cistern. The sculpture was fabricated, delivered, and installed by the project owner's contractor. The installation took place before site work such as the sidewalk and curbs were constructed, requiring the general contractor to meticulously coordinate the installation with other trade activities on-site. Because the sculpture was in an irregular shape, its hoisting and erection posed unique challenges to all stakeholders. Figure 10.6 shows how a mobile crane was employed to lift the sculpture into its designated location, with related field personnel standing close by on a ladder and a man lift to guide the erection. During the installation, the crane had to park on a public road (i.e., NE

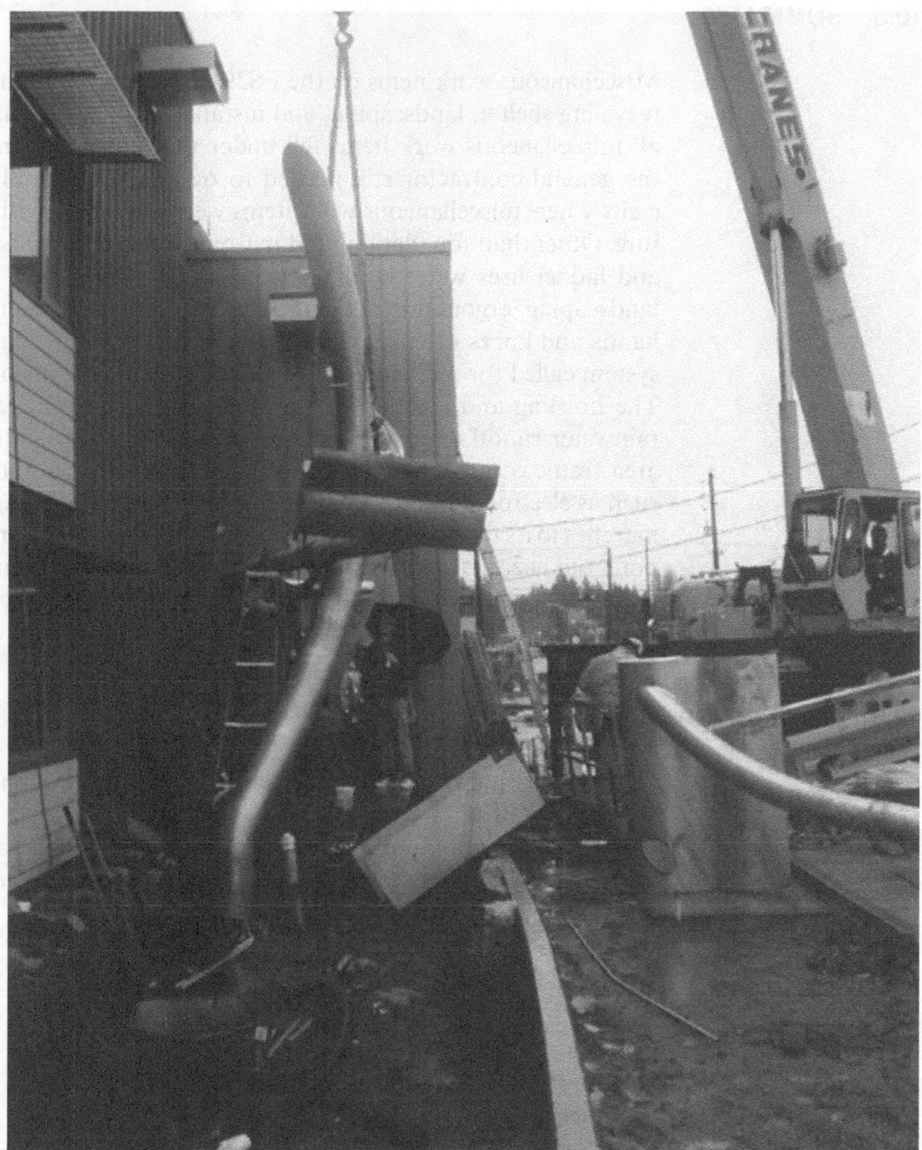

Figure 10.6 Steel art sculpture being lifted into its location on FS39
(Courtesy of The Miller Hull Partnership, LLP.)

127th St.), requiring the general contractor to limit the local area traffic to keep passing pedestrians and motor vehicles away from the hoisting operation.

The presence of a crane on-site exposed its surrounding workers to many potential hazards such as electrocution, struck-by (when someone is within a crane's swing radius), crane tipping, and hoisting failures. A detailed review of the safe operation of mobile cranes can be found in the discussion related to steel framing in Chapter 7.

10.5 SUMMARY

Miscellaneous work items on the FS39 project included the fueling and garbage recycling shelter, landscaping, and installation of a steel artwork. Although not all miscellaneous work items fell under the general contractor's direct control, the general contractor still needed to oversee the overall site safety and especially where miscellaneous work items were integrated with the completed structure. Other than foundation- and masonry-related hazards, falling from a height and ladder uses were the major concerns during shelter construction. During landscaping, ergonomic challenges surfaced due to workers often being on their hands and knees or engaging in heavy lifting. The installation of the irrigation system called for meticulous precaution to the locations of existing utility lines. The hoisting and erection of the steel sculpture for collecting and redirecting rainwater runoffs into the underground cistern highlighted the importance of area traffic control done by the general contractor as well as potential hazards such as electrocution, struck-by, crane tipping, and hoisting failures. As a project marches to its final completion, field personnel can easily drop their guard against potential hazards and constant training as well as communication are crucial to maintaining the safety culture all the way through a project's completion.

10.6 REVIEW QUESTIONS

1. What could have been done to ensure the safety of the worker illustrated in Figure 10.3?
2. Why would a fall restraint system be preferred over a fall arrest system for the worker engaging in the shelter roof construction at the FS39?
3. What is a stretch-and-flex program?
4. What strategies can be used to reduce the ergonomic challenges found during landscaping activities?

10.7 EXERCISES

1. Why would landscaping activity interfere with existing civil and electrical work on the FS39?
2. Did the installation of curbstones shown in Figure 10.5 create any safety concerns for the nearby activities? Please explain your answer.
3. Was the steel sculpture hoisting a critical lift? Why or why not?
4. Do musculoskeletal disorders occur often within construction? How do injuries caused by musculoskeletal disorders influence related stakeholders?

APPENDIX A
GLOSSARY

Accident prevention program (APP) a written safety and health program that establishes safety policies and procedures to be followed by all employees of the company.

Accredited testing organization private-sector organization that has been evaluated by OSHA for assurance that the organization meets the necessary qualifications to be considered as a nationally recognized testing laboratory.

Administrative controls changing how or when workers do their jobs, such as scheduling work or rotating workers to reduce exposure.

Alternative work a different job within a company that meets the physical restrictions imposed by a physician for an injured employee.

American National Standards Institute (ANSI) publisher of *American National Standards*, a reference book containing approved standards and specifications for building construction.

Anchorage a secure point of attachment for lifelines or lanyards.

Barricade an obstruction to deter the passage of persons or vehicles.

Base premium the workers' compensation premium paid based on the work classification of the employee before being modified by the experience modification ratio.

Building information modeling (BIM) digital, model-based technology that is linked to a database of project information.

Bureau of Labor Statistics (BLS) a unit of the U.S. Department of Labor that is the principal fact-finding agency for the federal government in the broad field of labor economics and statistics.

Code of Federal Regulations (CFR) a uniform system of listing all of the regulations promulgated by federal agencies.

Competent person a person designated by a construction company who is responsible for regular inspections of a project site for conformance with required safety practices and procedures.

Confined space an area which is enclosed with limited access and has the potential for a significant atmosphere hazard.

Controlled decking zone (CDZ) a work area that is clearly marked and designated in which certain types of work may take place without the use of conventional fall protection systems.

Critical lift the maximum load that can be lifted safely by a lifting device, such as a crane.

DART incidence rate the number of days away plus the number of days of restricted work plus the number of days of job transfer times 200,000 divided by the number of hours worked by all employees.

Emergency response plan a written document that delineates critical information specific to a project for timely response and communication in the event of unexpected emergencies.

Engineering controls physically changing a work environment.

Ergonomic designed to minimize physical effort and discomfort and hence maximize efficiency.

Experience modification ratio (EMR) a workers' compensation premium modifier used to account for the company's claims history during the earliest three of the preceding four years.

Globally harmonized system of classification and labeling of chemicals (GHS) an internationally agreed-upon system created by the United Nations to replace the various classification and labeling standards used in different countries by using consistent criteria for classification and labeling on a global level.

Hazard communication standards (HCS) a common and coherent approach to classifying chemicals and communicating hazard information on labels and safety data sheets.

Incidence rate a statistic for measuring safety performance that is determined by dividing the number of recordable injuries and illnesses by 200,000 hours.

Job hazard analysis (JHA) breaking down a specific construction activity into specific tasks, listing all potential hazards associated with each task, and identifying proper methods to eliminate or reduce the hazards to an acceptable risk level.

Lanyard a flexible line for connecting a body belt or harness to an anchorage.

Liability insurance insurance that provides coverage for financial loss that results from injuries or from property loss sustained by third parties as a consequence of a construction firm's activities.

Lock-out-tag-out (LOTO) procedure a safety policy that requires tags to be placed on the controls of all electrical controls of electrical circuits that are to be deactivated during the course of work on energized electrical equipment or circuits.

Material safety data sheets (MSDSs) forms published by manufacturers of hazardous materials that describe all known hazards associated with the materials and provide procedures for using, handling, and storing the materials safely.

Modified work an adjustment or alteration to the way in which a job is normally performed to accommodate the physical restrictions imposed by a physician on the working conditions for an injured employee.

Monopolistic state funds state workers' compensation funds from which employers must purchase insurance unless they choose to be self-insuring.

Musculoskeletal disorder health problems ranging from discomfort to minor aches and pains to more serious medical conditions that affect the body's muscles, joints, tendons, ligament, and nerves and are caused by the work being performed or the work environment.

National Fire Protection Association (NFPA) a trade association established to reduce the burden of fire and other hazards by creating and maintaining copyright standards and codes for usage and adoption by local governments, including the Fire Code, National Fuel Gas Code, National Electric, and Life Safety Code.

National Institute for Occupational Safety and Health (NIOSH) an agency within the U.S. Department of Health and Human Services that conducts research related to occupational safety and health and annually publishes a list of all known toxic substances.

Near-miss event an unplanned event that did not result in injury, illness, or damages but had the potential to do so.

Notice of contest an employer appeal to an OSHA citation.

Notice to Proceed written communication issued by a project owner to the construction contractor authorizing the contractor to proceed with the project and establishing the date of project commencement.

Occupational Safety and Health Act (OSH Act) a federal law passed in 1970 to assure, insofar as possible, safe and healthy working conditions for every working man and woman by establishing mandatory workplace safety and health procedures.

Occupational Safety and Health Administration (OSHA) an agency within the U.S. Department of Labor that is responsible for establishing safety and health standards, conducting inspections, issuing citations for violations of the standards, assessing penalties for noncompliance, providing safety training and injury prevention consultation, and maintaining a database of health and safety statistics.

OSHA Form 300 log of work-related injuries and illnesses

OSHA Form 300A summary of work-related injuries and illnesses.

OSHA Form 301 injury and illness incident report.

Part-time work working less than a normal work schedule.

Permanent partial disability a disability that, although permanent, does not completely limit a person's ability to work.

Permanent total disability a permanent disability that precludes all work.

Permissible exposure limit (PEL) an exposure limit that is published and enforced by OSHA as a legal standard.

Personal fall arrest system (PFAS) A system, consisting of an anchorage, connectors, and a body belt or harness, that controls the fall to a specified distance and limits the force to which an individual is subjected in the event of a fall.

Personal protective equipment (PPE) equipment or a device that protects a worker's body from hazards and any harmful conditions that may result in injury, illness, or possibly death.

Petition for modification of abatement an employer's contest of the time allocated to correct an OSHA violation.

Phased safety planning a systematic approach that examines the logical sequence of required construction activities and highlights the potential hazards associated with each activity.

Project-specific safety and health plan a safety and health plan developed for a specific project that identifies all hazards associated with each phase of the project and prescribes specific mitigation measures to be taken to minimize the potential harmful effect of each hazard.

Recordable injury and illness work-related death, injuries, and illnesses other than minor injuries requiring only first-aid treatment and which do not involve medical treatment, loss of consciousness, restriction of work or motion, or transfer to another job.

Respirator a device that is designed to protect the wearer from inhaling harmful atmospheres.

Retrospective workers' compensation insurance a safety incentive program in which a company pays a standard premium for its workers' compensation insurance, but the actual premium is increased or decreased at the end of a year based on the employer's claim history during that year.

Safety and health policy the first and most important part of a company's safety and health program defining management's commitment to a safe and healthful work environment.

Safety data sheets (SDSs) a new term for material safety data sheets.

Schedule disabilities a schedule for compensating permanent impairment based on body parts affected.

Severity rate the time lost through injuries as calculated in total days lost per 1000 hours worked.

Standard threshold shift (STS) a change in hearing threshold, relative to the baseline audiogram for an employee, of an average of 10 dB in one or both ears.

Tags temporary signs, usually attached to a piece of equipment or electrical controls, to warn of existing hazards.

Temporary Erosion and Sediment Control (TESC) Plan a plan to prevent erosion and the transport of sediment from a site during construction.

Temporary partial disability a temporary disability that does not completely limit a person's ability to work.

Temporary total disability a temporary disability that does not completely limit a person's ability to work.

Transitional job work accommodation to enable an injured employee to return to work when restrictions preclude performing the job held when the injury occurred.

Workers' compensation insurance a no-fault insurance in which employees give up the right to sue their employers for compensation resulting from injury or illness sustained in the workplace and employers provide compensation irrespective of whether or not the employee's negligence contributed to the injury.

APPENDIX B
FIRE STATION 39

SAMPLE ACCIDENT PREVENTION PROGRAM

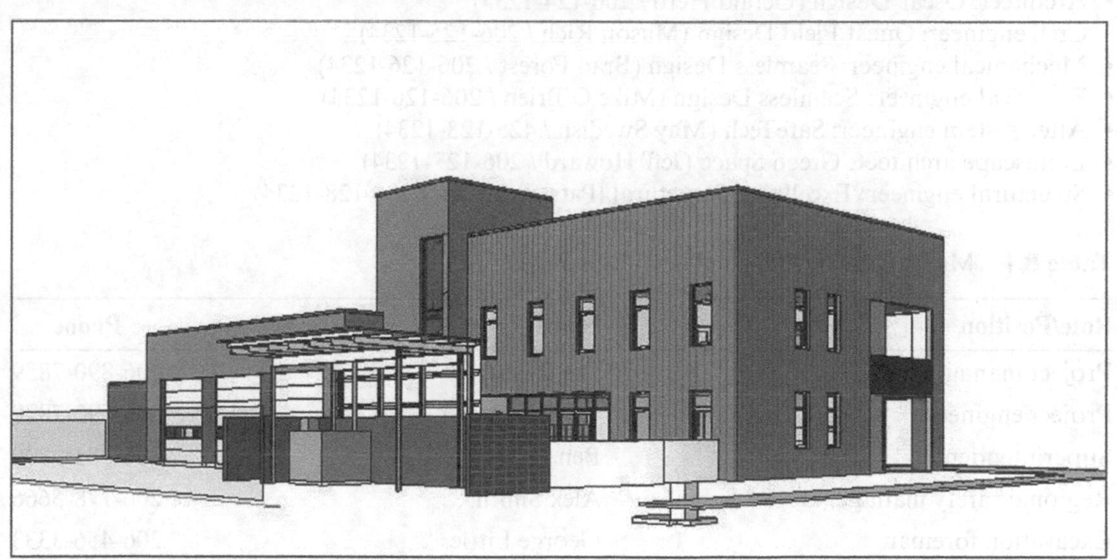

(All the company and person names in this plan are fictitious. Any resemblance to real business entities or real persons is purely coincidental)

2806 NE 137th Street
Seattle, WA 98025

(continued)

(*Continued*)

Project Information

- Name: Fire Station 39
- Address: 2806 NE 127th St., Seattle, WA 98125
- Phone: 206-980-1666
- Duration: April 2009–March 2010
- Description: The project is an 11,200-ft^2, structural steel frame building that includes a two-story station house, a one-story apparatus bay, and a one-story storage wing. Existing pavement on proposed project site will be removed entirely. Existing temporary structures will be relocated prior to construction. Finally, the project is to achieve the LEED Silver certification as per the City of Seattle's requirement.
- Owner: City of Seattle (MJ Simpson / 206-123-1234)
- Architect: Oscar Design (Gerald Hero / 206-124-1234)
- Civil engineer: Quest Field Design (Mason Rich / 206-125-1234)
- Mechanical engineer: Seamless Design (Brad Forest / 206-126-1234)
- Electrical engineer: Seamless Design (Mike O'Brien / 206-126-1234)
- Alter system engineer: SafeTech (May Swedish / 425-123-1234)
- Landscape architect: Green Space (Jeff Howard / 206-127-1234)
- Structural engineer: Excellence Structural (Patrick Brown / 206-128-1234)

Table B.1 Major Contact Information for the Project

Role/Position	Name	Phone
Project manager	Jon Rushford	206-890-7889
Project engineer	Charles Power	206-890-7878
Superintendent	Ben Dennis	206-785-9900
Regional safety manager	Alex Smith	206-778-5666
Excavation foreman	George Little	206-456-3333
Steel erection foreman	Joe Brothers	206-891-1125
Electrical foreman	Kyle Minch	206-456-4571
Plumbing/HVAC foreman	Eric Howe	206-359-1231
Locate	Call-Before-You-Dig	8-1-1

Our Commitment to Safety

Safety comes first. At CBE Construction, we believe that the company's biggest asset is our employees and every employee has the right to a safe and healthful work environment. The company is committed to the highest safety standards and practices, with the goal to reduce our injury rate by 10% each year. Compliance with the Accident Prevention Program and all related safety polices and rules is mandatory for everyone at CBE Construction. The responsibilities for site-specific leadership are described as follows.

The project manager (Jon Rushford) will take an active role in supporting CBE Construction's commitment to safety and health on-site. The project manager will lead the overall strategic planning and analysis for site safety, allocate necessary resources, and resolve conflicts.

The site superintendent (Ben Dennis) will ensure that the Accident Prevention Program is followed on-site and that all safety and health policies, rules, and regulations are observed in the field. The superintendent will also conduct daily job walks, take disciplinary actions when needed, organize weekly safety meetings, and lead accident investigations and reporting.

The regional safety manager (Alex Smith) will serve as the resource and go-to person when questions related to the legal requirements for health and safety arise. The safety manager will also perform weekly job walks, conduct employee trainings, assist the superintendent in accident investigations and reporting, and participate in meetings related to major safety issues on-site.

Project Administration

Discipline and Accountability

When the company's safety policies are violated, the following disciplinary steps will be followed:

1. A first-time violation will receive a verbal warning.
2. For second-time offenders, a written notification will become a part of the violator's official record.
3. A third violation will result in time suspension or possible employment termination, depending on the seriousness of the violation.
4. Employees who purposefully engage in misconducts or violations that could cause life-threatening situations or severe property damage will be terminated immediately.

Safety Inspection

Site safety inspections will occur at the beginning of the project and once a week afterwards by the regional safety manager (Alex Smith). Findings from the safety inspections will be discussed in the weekly safety and subcontractor coordination meetings on-site and in the monthly corporate safety meetings.

Accident Investigation

Accident investigation will be conducted by the project superintendent (Ben Dennis) with assistance from the regional safety director (Alex Smith). The purpose of the investigation is not to identify the liable parties but to find the root causes of the accidents for future prevention. Accidents, including personal injuries, exposure to blood, bodily fluids, or hazardous substance, property damages, and near-misses, regardless of their nature shall be reported to the project manager or superintendent for immediate attention. The project superintendent should use the company's form AC104 to file the investigation results. The "Record Keeping and Reporting" section describes the procedures for handling recordable and reportable incidents.

(continued)

(*Continued*)

Record Keeping and Reporting

All injuries, whether big or small, must be reported to the project superintendent (Ben Dennis). The superintendent must notify the regional safety manager (Alex Smith) and the human resources department for any injury that occurs on-site and file an internal investigation report using the company's form AC104 to describe the nature of the injury. The regional safety manager must update the injury and illness report (OSHA Form 301) and the log (OSHA Form 300) accordingly if the injury is recordable (anything beyond first aid). The summary (OSHA Form 300A) of the OSHA forms should be posted on-site for employee access together with other information essential for the workers' rights to a safe and healthful work environment. Should the incident involve any fatality or the hospitalization of one or more employees, the regional safety manager must notify (1) the regional executive officer immediately and (2) the Washington State Department of Labor and Industries within 8 hours of the incident.

Training

All employees, including those from the subcontractors, will be required to complete a site orientation safety training with the project manager (Jon Rushford), superintendent (Ben Dennis), or project engineer (Charles Power) prior to entering the job site. Safety orientation hard-hat stickers should be worn by those who successfully completed the training for verification purposes.

Other than the walk-through orientation, the regional safety director (Alex Smith) will work with the project superintendent (Ben Dennis) to identify specialty trainings that are required in lieu of the site-specific hazards listed below:

- Confined space
- Fall hazards

Weekly site safety meetings and daily tool box meetings will serve as additional training opportunities to raise the awareness of employees and reflect upon safety issues taking place on-site.

All training should be documented using the company's form TR01. Documentation of the training should be filed with the human resources department for a minimum of five years.

Medical Assistance and First Aid

Per CBE Construction policy, all project managers, superintendents, and foremen are first-aid and CPR certified. First-aid kits are located in the job trailer and all company vehicles. Additional kits available on-site are marked in the emergency response plan. If an employee requires medical attention beyond first aid and the condition is not life threatening, the employee must be escorted by the site superintendent (Ben Dennis), project engineer (Charles Power), or a designated employee to the nearby medical facilities (see the directions below). For severe injuries and illnesses, call 9-1-1 and then notify the superintendent immediately. The superintendent or a designated employee will assist the medical team at the site entrance (northwest corner of the site on 28th Ave. NE).

Figure B.1 Map to the nearest hospital for 24/7 ER and emergencies

Route: 2.9 mi, 9 min

A	**2806 NE 127th St, Seattle, WA 98125**	**A–B: 2.9 mi** 9 min
	1. Depart NE 127th St toward 30th Ave NE	**279 ft**
↱	**2. Turn right onto 30th Ave NE**	**0.2 mi**
↱	**3. Turn right onto WA-522 / Lake City Way NE**	**0.6 mi**
↗	**4. Bear right onto NE Northgate Way**	**1.4 mi**
↑	**5. Keep straight onto N Northgate Way**	**0.2 mi**
↱	**6. Turn right onto Meridian Ave N** *7-Eleven on the corner*	**0.3 mi**
↰	**7. Turn left onto N 115th St**	**0.2 mi**
B	**8. Arrive at 1550 N 115th St, Seattle, WA** *If you reach Stone Ave N, you've gone too far*	

Figure B.2 Directions to the nearest hospital for 24/7 ER and emergencies

- For 24/7 ER and emergencies: Northwest Hospital Medical Center (206-364-0500)
- For nonemergencies: Group Heath Northgate Medical Center (206-302-1200)

(continued)

(*Continued*)

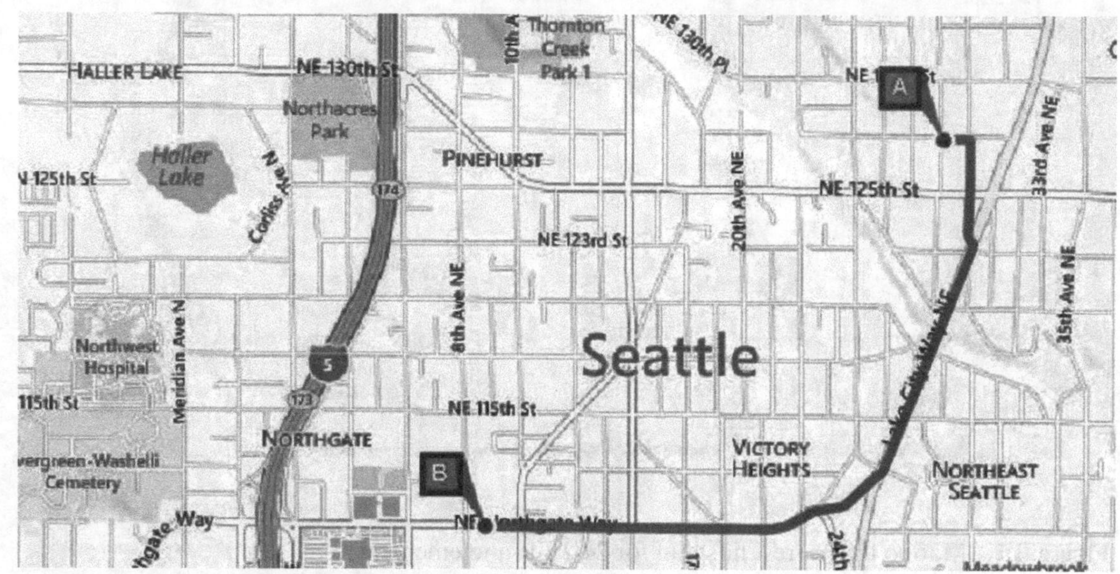

Figure B.3 Map to the nearest clinic for nonemergencies

Route: 1.8 mi, 5 min

A	**2806 NE 127th St, Seattle, WA 98125**	**A–B: 1.8 mi** 5 min
	1. Depart NE 127th St toward 30th Ave NE	**279 ft**
↱	**2. Turn right onto 30th Ave NE**	**0.2 mi**
↱	**3. Turn right onto WA-522 / Lake City Way NE**	**0.6 mi**
↱	**4. Bear right onto NE Northgate Way**	**0.9 mi**
B	**5. Arrive at 836 NE Northgate Way, Seattle, WA** *The last intersection is Roosevelt Way NE* *If you reach 8th Ave NE, you've gone too far*	

Figure B.4 Directions to the nearest clinic for nonemergencies

Any employee receiving medical attention due to work-related injuries or illnesses will be given an Employee Care Information Packet so that the employee can become familiar with the mandatory postaccident drug test program and the light-duty program.

Emergency Response Plan

In case of emergencies, call 9-1-1 and then notify the superintendent immediately. If the superintendent is not available, follow the line-of-communication plan and notify the next contact in charge. Alert other workers on-site and evacuate as needed to the designated gathering point. Foremen will conduct a head count for their crews and report back to the superintendent. In the case of personnel injury, follow the "Medical Assistance and First Aid" procedure.

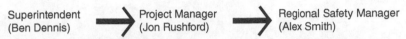

Superintendent (Ben Dennis) → Project Manager (Jon Rushford) → Regional Safety Manager (Alex Smith)

Figure B.5 Line of communication

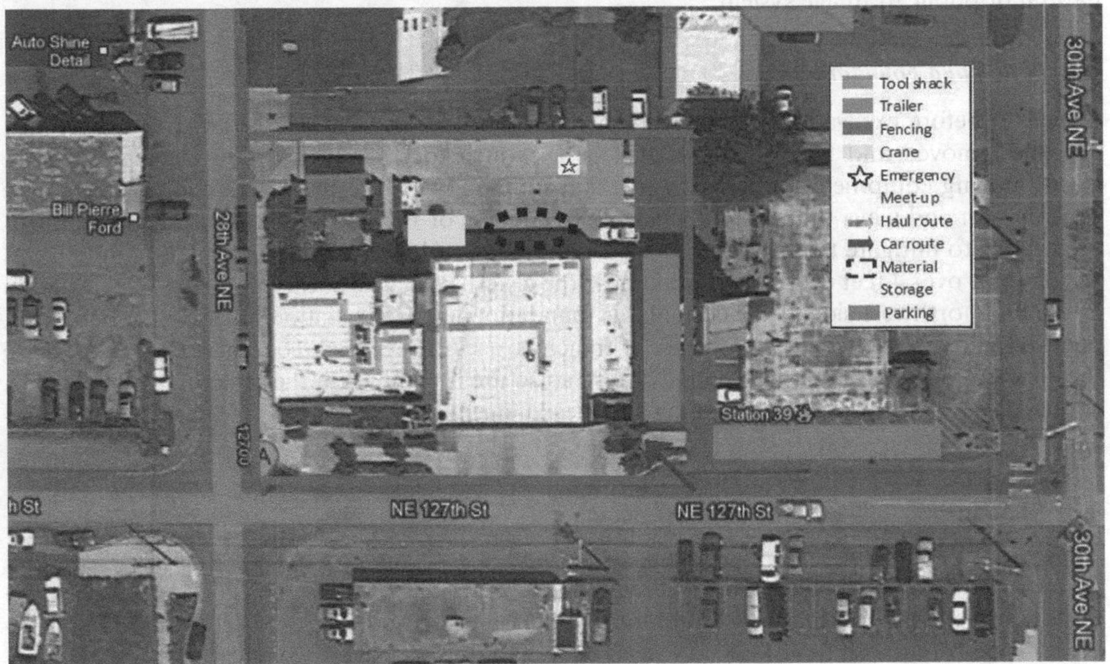

Figure B.6 Site logistics plan (showing the designated gathering point in case of emergencies)

Phase Project Planning: Site-Specific Hazards

Public Safety

The project is located within the boundary of the residential/commercial district with public service buildings nearby. Although the site is not right off the busy road (30th Ave. NE), light vehicle and foot traffic is still expected. This hazard can be mitigated by (1) the enclosure and fencing of sidewalks along both 28th Ave. NE and NE 127th St., (2) the use of flagger and proper signage, and (3) constant communication with the nearby businesses and residential community.

(continued)

(*Continued*)

Overhead Power Lines and Underground Utilities

Existing underground utilities are to be relocated by Seattle City Light at the project startup. Overhead power lines, about 40 feet above ground, run along both 28th Ave. NE and NE 127th St. across the streets and away from the site. A minimum clearance of 20 feet must be maintained between the power lines and any equipment used on-site.

Mobilization

The site is in a salmon watershed area. Temporary erosion and sedimentation collection facilities should be used as needed to ensure that sediment-laden water does not leave the site or enter the natural or public drainage system.

Excavation and Foundation

Two days before excavation breaks ground, request utility locating services. Existing pavement is to be removed and utility trenches will be excavated together with future footing locations. Earth moving equipment and hauling trucks form the major hazard. A well-coordinated traffic control plan and ways to increase employees' awareness of their surrounding work environment are needed to mitigate the hazard. Employees should wear high-visibility vests at all times. Excavation at or over 4 feet deep (foundation near the cistern structure at the south end of the station house and northwest side of the continuous foundation under the station house) must be sloped, benched, shored, or protected by the use of trench boxes. Any exposed rebar must be capped. Once the concrete work for the underground cistern structure is completed, the structure will become a confined space and site-specific training and measurements will be required (see applicable technical sections).

CMU Construction and Structural Steel

CMU walls will be put up for the apparatus bay and will require an extensive amount of scaffold and a material handling hydraulic lift. Bracing the CMU walls is a must, considering the potential gusts that could pick up in the northwest during the winter season. The establishment of a limited-access zone is required for CMU walls over 8 feet tall (see applicable technical sections). Lots of steel erection work will take place after the CMU walls are up and will require the use of a mobile crane for material hoisting. Activities involving the use of a mobile crane are to be marked in the project schedule and their related JHA will be required prior to the start of these activities. Site storage coordination and communication will be reinforced to avoid conflicts of the storage space. The steel subcontractor will be required to submit a site-specific steel erection and safety plan as well as task-specific JHAs before any steelwork can start. Fall from heights and struck-by by overhead loads are the two major hazards that should be mitigated (see applicable technical sections).

Exterior Enclosure

Framing, sheathing, and glazing will take place with heavy use of ladders and man lift. Fall from heights is the biggest hazard (see applicable technical sections).

Interior Enclosure

Trade coordination in the sequencing of activities and lay-down area requirements will be tightened to avoid spatial conflicts. Ladders, rolling scaffolds, and stilts will be used for interior work and fall from heights will continue to be the main hazard. Some areas of the site might need additional task lighting and ventilation. Housekeeping, tool safety, and overhead protection (once the MEP rough-in is completed) should also be emphasized to all employees on-site.

Technical Sections

Excavation Safety

Call the utility locating services two days before excavating or digging. Protect excavations and trenches by shoring, benching, and/or sloping. Protect employees working in trenches with the use of trench boxes. Provide proper railing for ramps crossing excavations or trenches that are 4 feet deep or more.

Confined Space

Test the air quality before employees enter a confined space. Adequate ventilation or fresh air must be provided when employees work within the confined space and continuous monitoring of the air quality is also a must.

Concrete Forming and Placement

Protect all exposed rebar ends. Pull all nails from boards and formwork. Preplan the paths of concrete trucks and placement. Wear safety glasses and gloves during concrete placement. Use other PPE as needed.

Electrical Safety

Only qualified electricians are allowed in power panels or to work on temporary power boxes. Ground fault circuit interrupters (GFCIs) are required on-site.

Scaffold Safety

Guardrails (top rails, midrails, and toe boards) must be installed on all open sides of scaffolds more than 6 feet in height, as the company has a stricter policy on scaffold safety. Scaffold planks must extend over the end supports at least 6 inches but no more than 12 inches. The scaffold must be on rigid footing, have adequate means of access, and be fully planked at the work level. Rolling scaffolds must have brakes that are locked prior to climbing or working on scaffold. See manufacturer's recommendations for outrigger use and installation.

Fall Protection

A written fall protection work plan shall be developed before the start of work for sites that anticipate fall hazards that are 10 feet or more. For the FS39 project, a fall protection work plan will be required by any subcontractor who needs to access the roof. A full body harness and lanyard

(*continued*)

(*Continued*)

assembly shall be used when employees are exposed to a fall hazard greater than 6 feet and are not protected by a fall restraint system. Open-sided floor, platform, or surfaces 4 feet or more above the adjacent floor or ground level shall be guarded by a standard railing or equivalent. Holes and openings shall be guarded or covered.

Ladder Safety

All ladders should be in good condition and without damage. Step ladders shall be used only in the full open position and shall not be used as an extension ladder. Non—step ladders should be secured and extend 3 feet above the upper level. Maintain three points of contact while ascending or descending a ladder.

Power Aerial Work Platforms

No employees are allowed to use or operate the platforms before completion of proper training. A fall arrest harness and lanyard is required for boom-supported elevated work platforms.

PPE

Hard hats, safety glasses, high-visibility clothing, and proper protective foot wear are mandatory. Other PPEs are provided to site employees as needed.

Tool Safety

Inspect tools daily and remove defective tools immediately for disposal or repair. Proper guards must be in place for power saws, grinders, and other power tools. Never yank or hoist a power tool by its cord. Keep tool cords and hoses out of walkways and away from sharp edges. All portable electrical tools must be grounded unless they are double insulated and approved. All fuel-powered tools must be shut down while being refueled and smoking is prohibited during refueling. Only employees who have valid credentials are permitted to use powder-actuated tools.

HazCom

A list of all known hazardous chemicals on-site and their material safety data sheets will be kept in the project trailer office. Employees must be given proper training on the safe handling and emergency procedures for the listed chemicals. Containers carrying the chemicals must be properly labeled.

Fire Safety

All fire extinguishers shall be inspected by a competent person and properly maintained. All defective equipment shall be immediately removed from the job site and replaced with properly working equipment.

Housekeeping

All work areas will be cleaned up during the task and after the shift. All aisle ways, stairways, and landings will be kept free of debris.

Burning and Welding

Inspect work areas to einsure that no burning or welding sparks will fall on combustible materials on-site. Fire extinguishing equipment should be nearby and employees must wear eye protection and protective gloves. A 30-minute fire watch will be required after welding or burning is complete.

Fuel Storage

Oxygen and fuel gas cylinders must be separated from each other by at least 20 feet or by a 5-foot-tall barrier which has a 30-minute fire rating. There will be no smoking within 50 feet of fuel storage.

Table B.2 Emergency Contact Information for Related Agencies and Utility Companies

Contact Name	Contact Phone Number
Electricity (Seattle City Light)	206-684-3000
Gas (Puget Sound Energy)	1-888-225-5773
Water (Seattle Public Utilities)	206-386-1800 (for 24/7)
Washington Poison Center	1-800-222-1222
Seattle Police Department	206-625-5011
Seattle Fire Department Dispatcher	206-386-1495
Washington Department of L&I	1-800-547-8367
OSHA	1-800-321-OSHA (6742)

APPENDIX C
PROJECT SCHEDULE FOR FIRE STATION 39

Act ID	Description	Orig Dur	Early Start	Early Finish	Total Float
PROJECT SUMMARY					
0100	Bid Day	0	11MAR09 A		
0110	Analyze Bids & Award Project	5	11MAR09 A	17MAR09 A	
0150	Buy Out Project	20	18MAR09 A	14APR09 A	
0200	Notice to Proceed	0	16APR09 A		
0210	Pre Construction Meeting	7	16APR09 A	24APR09 A	
0250	Start Work in Field	0	27APR09 A		
0300	Construction Duration	205 *	27APR09 A	15FEB10	0
0310	Contract duration	252 *	16APR09 A	12APR10	13d
0470	Foundations	72 *	07MAY09 A	17AUG09 A	
0480	Structure	17 *	14JUL09 A	05AUG09 A	
0490	Exterior Closure	138 *	30JUL09 A	15FEB10	0
0500	1st Floor finishes	110 *	07SEP09 A	12FEB10	1d
0510	2nd Floor Finishes	107 *	14SEP09 A	15FEB10	0
0520	Site Work	180 *	29MAY09 A	11FEB10	55d
0530	Permanent Power	0		29JAN10	3d
0550	Performance Commissioning	10	21JAN10 A	01FEB10	3d
0560	Life Safety Systems Testing	5	26JAN10 A	03FEB10	3d
0570	Final Inspections	5	27JAN10 A	10FEB10	3d
0590	Substantial Completion	0		15FEB10	0
0595	Contract Substantial Completion	0		15MAR10	0
0600	LEED Exhaust building/Closeout	10	16FEB10	01MAR10	13d
0610	Closeout	30	02MAR10	12APR10	13d
0620	Final Completion	0		12APR10	13d
0630	Contract Final Completion	0		14APR10	0
0640	Contract Final Acceptance	0		29APR10	0
OWNER AWARD PROCESS					
0810	Pre-award process	12	11MAR09 A	26MAR09 A	
0820	Contract Award	0	27MAR09 A		
0830	Procure bonds/Insurance	10	27MAR09 A	09APR09 A	
0840	Owner execute award	4	10APR09 A	15APR09 A	
SITE PREPARATION					
1010	Mobilize on Site	5	27APR09 A	01MAY09 A	
1020	Fence & Secure Site	3	20APR09 A	20APR09 A	
1030	Erosion Control Mesures	5	05MAY09 A	06MAY09 A	
1035	Relocate Existing Power	5	01JUN09 A	03JUN09 A	
1040	Clear & Grade	5	07MAY09 A	26OCT09 A	

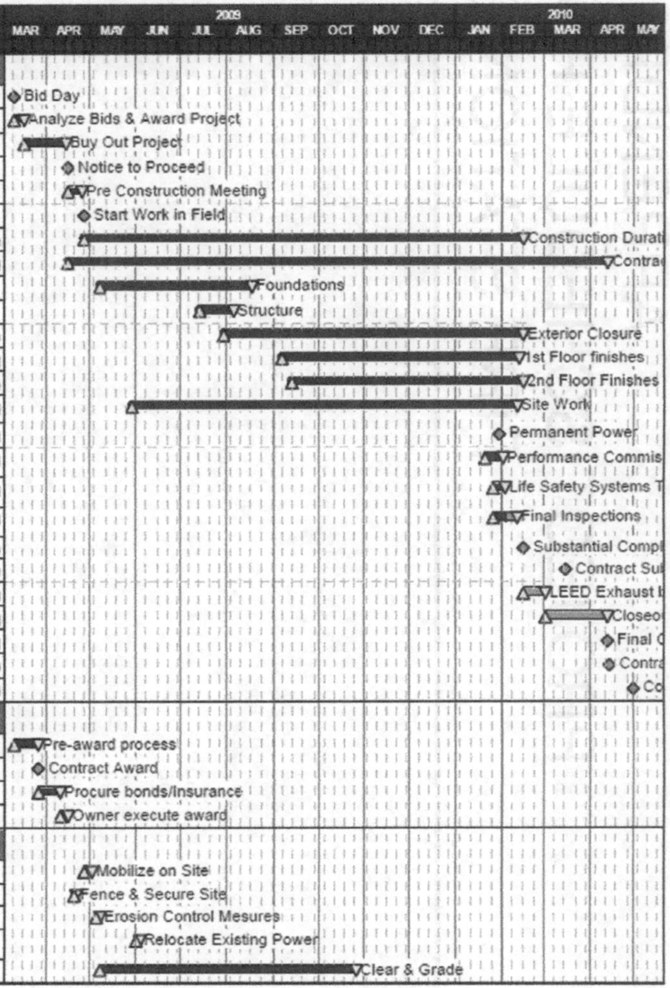

Figure C.1 Schedule summary, owner award process, and site preparation

Act ID	Description	Orig Dur	Early Start	Early Finish	Total Float
FOUNDATIONS					
2000	Structural Excavation	8	07MAY09 A	19MAY09 A	
2020	F/R/P Footings	12	08MAY09 A	28MAY09 A	
2040	F/R/P Foundation Walls	12	29MAY09 A	02JUL09 A	
2050	FRP Brace Frame Pilasters	5	06JUL09 A	06JUL09 A	
2060	FRP Cistern Walls	3	29MAY09 A	12JUN09 A	
2080	Build CMU Perimeter Walls	20	01JUN09 A	30JUN09 A	
2090	Underslab rough in	15	26MAY09 A	09JUL09 A	
2100	Trench Drains Apparatus Bay	5	29JUL09 A	03AUG09 A	
2200	Backfill, Capillary Break	5	29JUN09 A	10JUL09 A	
2330	F/R/P SOG	10	03AUG09 A	17AUG09 A	
STRUCTURE					
2410	Erect Bolt Weld Structural Steel	5	14JUL09 A	22JUL09 A	
2420	Erect Stairs	3	31JUL09 A	14AUG09 A	
2470	Metal Deck, Edge Forms & Studs	5	27JUL09 A	03AUG09 A	
2480	R/P SOMD Level 2	5	05AUG09 A	07AUG09 A	
2490	Fireproof Structural Steel Level 1	2	05AUG09 A	05AUG09 A	
EXTERIOR CLOSURE					
2700	Ext Wood Stud Framing & Sheathing	20	30JUL09 A	18SEP09 A	
2710	Roofing	15	27AUG09 A	30SEP09 A	
2720	Metal Siding, Extruded Aluminum & Coping	20	28SEP09 A	30DEC09 A	
2740	Cedar & Cement Fiber Siding	10	19OCT09 A	22JAN10 A	
2750	Paint Exterior CMU	10	18JAN10 A	21JAN10 A	
2752	Release of Sectional Doors	1	24NOV09 A	21JAN10 A	
2755	Sectional Doors	10	25JAN10 A	03FEB10	0
2757	Finalize Nederman System	1	28JAN10 A	08FEB10	0
2760	Storefront Windows	10	05OCT09 A	06JAN10 A	
2770	Punchlist	5	09FEB10	15FEB10	0
1st FLOOR FINISHES					
3500	MEP Rough in L1	15	07SEP09 A	16OCT09 A	
3510	Sand floors 1st pass	5	28SEP09 A	19OCT09 A	
3520	Frame Walls & Ceiling L1	15	12OCT09 A	26OCT09 A	
3530	Door Jambs L1	10	13OCT09 A	22OCT09 A	
3540	Wall & Ceiling Rough in L1	15	26OCT09 A	25NOV09 A	
3570	Insulate, GWB & Taping L1	15	20NOV09 A	02DEC09 A	
3580	Wood Ceiling & Walls L1	10	03DEC09 A	10DEC09 A	
3590	Prime , Paint Walls & Ceilings L1	10	08DEC09 A	11DEC09 A	
3600	MEP Trim L1	10	14DEC09 A	28JAN10 A	

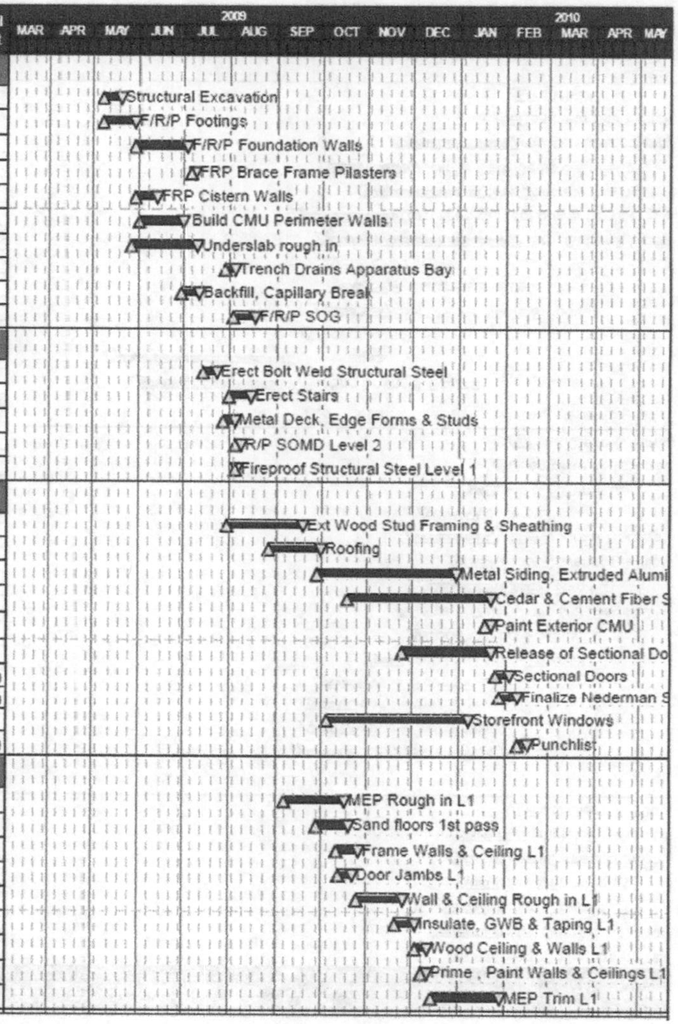

Figure C.2 Foundations, structure, exterior closure, and first-floor finishes

Act ID	Description	Orig Dur	Early Start	Early Finish	Total Float
3630	Floor Coverings & Wall Base L1	10	28DEC09 A	26JAN10 A	
3640	Install Doors & Hardware L1	10	16DEC09 A	22JAN10 A	
3650	Kitchen Cabinets L1	5	21DEC09 A	15JAN10 A	
3700	Lavs & Fixtures L1	5	04JAN10 A	21JAN10 A	
3710	Install wheel chair lift	15	28DEC09 A	01FEB10	5d
3720	CompleteFinishes L1	10	28DEC09 A	05FEB10	1d
3730	Final Clean & Punch	5	08FEB10	12FEB10	1d
2nd FLOOR FINISHES					
4510	MEP Overhead Rough in L2	15	14SEP09 A	16OCT09 A	
4520	Frame Walls & ceiling L2	15	21SEP09 A	16OCT09 A	
4530	Door Jambs L2	10	21SEP09 A	15OCT09 A	
4550	Wall Rough in L2	15	30SEP09 A	30OCT09 A	
4560	Insulate, GWB & Taping L2	15	02NOV09 A	19NOV09 A	
4570	Wood Ceiling & Walls L2	10	01DEC09 A	02DEC09 A	
4600	Prime & Paint Walls & Ceilings L2	10	24NOV09 A	24NOV09 A	
4640	MEP Trim P2	10	07DEC09 A	01FEB10	2d
4650	Floor Coverings & Wall Base L2	10	28DEC09 A	26JAN10 A	
4660	Install Doors & Hardware L2	5	30NOV09 A	22JAN10 A	
4670	Lockers L2	5	08DEC09 A	08JAN10 A	
4680	Lavs & Fixtures L2	5	05JAN10 A	18JAN10 A	
4700	Complete Finishes L2	10	25JAN10 A	04FEB10	2d
4710	Final Clean & Punch	5	05FEB10	11FEB10	2d
SITEWORK					
4750	Footing Drains & Backfill	10	29MAY09 A	14SEP09 A	
4760	Site Utilities	15	12MAY09 A	11SEP09 A	
4770	Rough Grade Site	10	21SEP09 A	25SEP09 A	
4780	Fuel Shelter Walls & Footings	5	28SEP09 A	11NOV09 A	
4790	FRP Concrete Sitewalls	10	06OCT09 A	28OCT09 A	
4800	Site CMU Walls	10	05OCT09 A	09OCT09 A	
4810	Curb, Gutter & Sidewalks	10	30SEP09 A	05FEB10	59d
4820	Install permanent power service	30	14SEP09 A	23NOV09 A	
4870	Site Concrete Paving	20	27OCT09 A	15JAN10 A	
4890	Granite Curb Seats & Curbing	15	01OCT09 A	30NOV09 A	
4940	Irrigation Landscaping	20	18NOV09 A	02FEB10	55d
4950	Complete Site Finishes	10	26JAN10 A	04FEB10	55d
4960	Punchlist	5	05FEB10	11FEB10	55d

Figure C.3 Second-floor finishes and site work

APPENDIX D
OSHA 10/30 TOPIC COVERAGE

Table D.1 OSHA 10 Topic Coverage

10-hr Topics	Required or Elective	Corresponding Chapter
Introduction to OSHA	Required	Chapter 4
OSHA Focus Four Hazards—Fall Protection	Required	Chapter 7, Chapter 8, Chapter 9, Chapter 10, Appendix A
OSHA Focus Four Hazards—Electrical	Required	Chapter 7, Appendix B
OSHA Focus Four Hazards—Struck By	Required	Chapter 7, Appendix B
OSHA Focus Four Hazards—Caught in/ Between	Required	Chapter 6, Appendix B
Personal Protective and Lifesaving Equipment	Elective	Chapter 7, Chapter 9, Appendix B
Health Hazards in Construction	Elective	Chapter 3, Appendix B
Material Handling, Storage, Use and Disposal	Elective	Chapter 7, Appendix B
Tools—Hand and Power	Elective	Chapter 8, Chapter 9, Appendix B
Scaffold	Elective	Chapter 7, Appendix B
Cranes, Derricks, Hoists, Elevators, and Conveyors	Elective	Chapter 7
Excavations	Elective	Chapter 6, Appendix B
Stairways and Ladders	Elective	Chapter 9, Appendix B

Table D.2 OSHA 30 Topic Coverage

30-hr Topics	Required or Elective	Corresponding Chapter
Introduction to OSHA	Required	Chapter 4
OSHA Focus Four Hazards—Fall Protection	Required	Chapter 7, Chapter 8, Chapter 9, Chapter 10, Appendix A
OSHA Focus Four Hazards—Electrical	Required	Chapter 7, Appendix B
OSHA Focus Four Hazards—Struck by	Required	Chapter 7, Appendix B
OSHA Focus Four Hazards—Caught in/Between	Required	Chapter 6, Appendix B
Personal Protective and Lifesaving Equipment	Required	Chapter 7, Chapter 9, Appendix B
Health Hazards in Construction	Required	Chapter 3, Appendix B
Stairway and Ladders	Required	Chapter 9, Appendix B
Managing Safety and Health	Required	Chapter 2, Chapter 3, Chapter 4
Fire Protection and Prevention	Elective	Appendix B
Material Handling, Storage, Use and Disposal	Elective	Chapter 7, Appendix B
Tools—Hand and Power	Elective	Chapter 8, Chapter 9, Appendix B
Welding and Cutting	Elective	(—)
Scaffolding	Elective	Chapter 7
Cranes, Derricks, Hoists, Elevators, and Conveyors	Elective	Chapter 7
Motor Vehicles, Mechanized Equipment and Marine Operations; Rollover Protective Structure and Overhead Protection; and Signs, Signals, and Barricades	Elective	Chapter 7
Excavations	Elective	Chapter 6
Concrete and Masonry Construction	Elective	Chapter 7
Steel Erection	Elective	Chapter 7
Safety and Health Program	Elective	Chapter 3, Appendix B
Confined Space Entry	Elective	Chapter 6
Power Industrial Vehicles	Elective	(—)
Ergonomics	Elective	Chapter 10

APPENDIX E
SELECTED OSHA REQUIREMENTS

This appendix contains abridged versions of most of the OSHA standards* that relate to construction projects. While many states have published their own occupational safety and health requirements, such requirements generally are based on OSHA standards. For this reason, the authors decided to use the basic OSHA requirements as the context for the discussion presented in the book and for the development of safety plans for Fire Station 39. The requirements contained in this appendix can also be used by students in developing project-specific safety plans for other construction projects. It is very important that students learn to use these requirements when developing safety plans for construction projects.

1926 SUBPART C—GENERAL SAFETY AND HEALTH PROVISIONS

This part of the OSHA standards is fundamental to any construction project and is discussed throughout the book. Specifically, many of the requirements here are applied in the sample Fire Station 39 Accident Prevention Program in Appendix C, and Chapter 9 discusses the application of the housekeeping and illumination requirements.

1926.20—General Safety and Health Provisions

(a) Contractor requirements. Section 107 of the Act requires that it shall be a condition of each contract which is entered into for construction, alteration, and/or repair, including painting and decorating, that no contractor or subcontractor for any part of the contract work shall require any laborer or mechanic employed in the performance of the contract to work in surroundings or under working conditions which are unsanitary, hazardous, or dangerous to his health or safety.

(b) Accident prevention responsibilities.

(1) It shall be the responsibility of the employer to initiate and maintain such programs as may be necessary to comply with this part.

(2) Such programs shall provide for frequent and regular inspections of the job sites, materials, and equipment to be made by competent persons designated by the employers.

(*The contents of this appendix are based on the OSHA requirements as of May 1, 2013, without the subpart appendices. The requirements were cited from http://www.osha.gov/)

(3) The use of any machinery, tool, material, or equipment which is not in compliance with any applicable requirement of this part is prohibited. Such machine, tool, material, or equipment shall either be identified as unsafe by tagging or locking the controls to render them inoperable or shall be physically removed from its place of operation.

(4) The employer shall permit only those employees qualified by training or experience to operate equipment and machinery.

(c) The standards contained in this part shall apply with respect to employments performed in a workplace in a State, the District of Columbia, the Commonwealth of Puerto Rico, the Virgin Islands, American Samoa, Guam, Trust Territory of the Pacific Islands, Wake Island, Outer Continental Shelf lands defined in the Outer Continental Shelf Lands Act, Johnston Island, and the Canal Zone.

(d) If a particular standard is specifically applicable to a condition, practice, means, method, operation, or process, it shall prevail over any different general standard which might otherwise be applicable to the same condition, practice, means, method, operation, or process. On the other hand, any standard shall apply according to its terms to any employment and place of employment in any industry, even though particular standards are also prescribed for the industry to the extent that none of such particular standards applies.

(e) In the event a standard protects on its face a class of persons larger than employees, the standard shall be applicable under this part only to employees and their employment and places of employment.

(f) Compliance duties owed to each employee.

(1) Personal protective equipment. Standards in this part requiring the employer to provide personal protective equipment (PPE), including respirators and other types of PPE, because of hazards to employees impose a separate compliance duty with respect to each employee covered by the requirement. The employer must provide PPE to each employee required to use the PPE, and each failure to provide PPE to an employee may be considered a separate violation.

(2) Training. Standards in this part requiring training on hazards and related matters, such as standards requiring that employees receive training or that the employer train employees, provide training to employees, or institute or implement a training program, impose a separate compliance duty with respect to each employee covered by the requirement. The employer must train each affected employee in the manner required by the standard, and each failure to train an employee may be considered a separate violation.

1926.21—Safety Training and Education

(a) General requirements. The Secretary shall, pursuant to section 107(f) of the Act, establish and supervise programs for the education and training of employers and employees in the recognition, avoidance and prevention of unsafe conditions in employments covered by the act.

(b) Employer responsibility.

(1) The employer should avail himself of the safety and health training programs the Secretary provides.

(2) The employer shall instruct each employee in the recognition and avoidance of unsafe conditions and the regulations applicable to his work environment to control or eliminate any hazards or other exposure to illness or injury.

(3) Employees required to handle or use poisons, caustics, and other harmful substances shall be instructed regarding the safe handling and use, and be made aware of the potential hazards, personal hygiene, and personal protective measures required.

(4) In job site areas where harmful plants or animals are present, employees who may be exposed shall be instructed regarding the potential hazards, and how to avoid injury, and the first aid procedures to be used in the event of injury.

(5) Employees required to handle or use flammable liquids, gases, or toxic materials shall be instructed in the safe handling and use of these materials and made aware of the specific requirements contained in Subparts D, F, and other applicable subparts of this part.

(6) (i) All employees required to enter into confined or enclosed spaces shall be instructed as to the nature of the hazards involved, the necessary precautions to be taken, and in the use of protective and emergency equipment required. The employer shall comply with any specific regulations that apply to work in dangerous or potentially dangerous areas.

(ii) For purposes of paragraph (b)(6)(i) of this section, "confined or enclosed space" means any space having a limited means of egress, which is subject to the accumulation of toxic or flammable contaminants or has an oxygen deficient atmosphere. Confined or enclosed spaces include, but are not limited to, storage tanks, process vessels, bins, boilers, ventilation or exhaust ducts, sewers, underground utility vaults, tunnels, pipelines, and open top spaces more than 4 feet in depth such as pits, tubs, vaults, and vessels.

1926.23—First Aid and Medical Attention

First aid services and provisions for medical care shall be made available by the employer for every employee covered by these regulations. Regulations prescribing specific requirements for first aid, medical attention, and emergency facilities are contained in Subpart D of this part.

1926.24—Fire Protection and Prevention

The employer shall be responsible for the development and maintenance of an effective fire protection and prevention program at the job site throughout all phases of the construction, repair, alteration, or demolition work. The employer shall ensure the availability of the fire protection and suppression equipment required by Subpart F of this part.

1926.25—Housekeeping

(a) During the course of construction, alteration, or repairs, form and scrap lumber with protruding nails, and all other debris, shall be kept cleared from work areas, passageways, and stairs, in and around buildings or other structures.

(b) Combustible scrap and debris shall be removed at regular intervals during the course of construction. Safe means shall be provided to facilitate such removal.

(c) Containers shall be provided for the collection and separation of waste, trash, oily and used rags, and other refuse. Containers used for garbage and other oily, flammable, or hazardous wastes, such as caustics, acids, harmful dusts, etc. shall be equipped with covers. Garbage and other waste shall be disposed of at frequent and regular intervals.

1926.26—Illumination

Construction areas, aisles, stairs, ramps, runways, corridors, offices, shops, and storage areas where work is in progress shall be lighted with either natural or artificial illumination. The minimum illumination requirements for work areas are contained in Subpart D of this part.

1926.27—Sanitation

Health and sanitation requirements for drinking water are contained in Subpart D of this part.

1926.28—Personal Protective Equipment

(a) The employer is responsible for requiring the wearing of appropriate personal protective equipment in all operations where there is an exposure to hazardous conditions or where this part indicates the need for using such equipment to reduce the hazards to the employees.

(b) Regulations governing the use, selection, and maintenance of personal protective and lifesaving equipment are described under Subpart E of this part.

1926.34—Means of Egress

(a) In every building or structure exits shall be so arranged and maintained as to provide free and unobstructed egress from all parts of the building or structure at all times when it is occupied. No lock or fastening to prevent free escape from the inside of any building shall be installed except in mental, penal, or corrective institutions where supervisory personnel is continually on duty and effective provisions are made to remove occupants in case of fire or other emergency.

(b) Exits shall be marked by a readily visible sign. Access to exits shall be marked by readily visible signs in all cases where the exit or way to reach it is not immediately visible to the occupants.

(c) Means of egress shall be continually maintained free of all obstructions or impediments to full instant use in the case of fire or other emergency.

1926.35—Employee Emergency Action Plans

(a) This section applies to all emergency action plans required by a particular OSHA standard. The emergency action plan shall be in writing (except as provided in the last sentence of paragraph (e)(3) of this section) and shall cover those designated actions employers and employees must take to ensure employee safety from fire and other emergencies.

(b) The following elements, at a minimum, shall be included in the plan:
  (1) Emergency escape procedures and emergency escape route assignments;
  (2) Procedures to be followed by employees who remain to operate critical plant operations before they evacuate;
  (3) Procedures to account for all employees after emergency evacuation has been completed;
  (4) Rescue and medical duties for those employees who are to perform them;
  (5) The preferred means of reporting fires and other emergencies; and
  (6) Names or regular job titles of persons or departments who can be contacted for further information or explanation of duties under the plan.

(c) "Alarm system."
  (1) The employer shall establish an employee alarm system which complies with 1926.159.
  (2) If the employee alarm system is used for alerting fire brigade members, or for other purposes, a distinctive signal for each purpose shall be used.

(d) The employer shall establish in the emergency action plan the types of evacuation to be used in emergency circumstances.

(e) "Training."
  (1) Before implementing the emergency action plan, the employer shall designate and train a sufficient number of persons to assist in the safe and orderly emergency evacuation of employees.
  (2) The employer shall review the plan with each employee covered by the plan at the following times;
    (i) Initially when the plan is developed,
    (ii) Whenever the employee's responsibilities or designated actions under the plan change, and
    (iii) Whenever the plan is changed.
  (3) The employer shall review with each employee upon initial assignment those parts of the plan which the employee must know to protect the employee in the event of an emergency. The written plan shall be kept at the workplace and made available for employee review. For those employers with 10 or fewer employees the plan may be communicated orally to employees and the employer need not maintain a written plan.

1926 SUBPART D—OCCUPATIONAL HEALTH AND ENVIRONMENTAL CONTROLS

1926.50—Medical Services and First Aid

The sample accident prevention program in Appendix C illustrates how this part of the OSHA standards can be addressed through project planning.

(a) The employer shall insure the availability of medical personnel for advice and consultation on matters of occupational health.

(b) Provisions shall be made prior to commencement of the project for prompt medical attention in case of serious injury.

(c) In the absence of an infirmary, clinic, hospital, or physician, that is reasonably accessible in terms of time and distance to the worksite, which is available for the treatment of injured employees, a person who has a valid certificate in first-aid training shall be available at the worksite to render first aid.

(d) First aid supplies shall be easily accessible when required. The contents of the first aid kit shall be placed in a weatherproof container with individual sealed packages for each type of item, and shall be checked by the employer before being sent out on each job and at least weekly on each job to ensure that the expended items are replaced.

(e) Proper equipment for prompt transportation of the injured person to a physician or hospital, or a communication system for contacting necessary ambulance service, shall be provided.

(f) In areas where 911 is not available, the telephone numbers of the physicians, hospitals, or ambulances shall be conspicuously posted.

(g) Where the eyes or body of any person may be exposed to injurious corrosive materials, suitable facilities for quick drenching or flushing of the eyes and body shall be provided within the work area for immediate emergency use.

1926.52—Occupational Noise Exposure

The need to have hearing protection is discussed in an example job hazard analysis for site preparation and grading in Chapter 3. At the Fire Station 39 project, during the masonry wall construction, power tools such as masonry saws would create noises and workers should be aware of how different noise exposure levels require proper protection.

(a) Protection against the effects of noise exposure shall be provided when the sound levels exceed those shown in Table D-2 of this section when measured on the A-scale of a standard sound level meter at slow response.

(b) When employees are subjected to sound levels exceeding those listed in Table D-2 of this section, feasible administrative or engineering controls shall be utilized. If such controls fail to reduce sound levels within the levels of the table, personal protective equipment as required in Subpart E, shall be provided and used to reduce sound levels within the levels of the table.

(c) If the variations in noise level involve maxima at intervals of 1 second or less, it is to be considered continuous.

Table D-2 Permissible Noise Exposures

Duration Per Day, hr	Sound Level dBA Slow Responses
8	90
6	92
4	95
3	97
2	100
1½	102
1	105
½	110
¼ or less	115

(d) (1) In all cases where the sound levels exceed the values shown herein, a continuing, effective hearing conservation program shall be administered.

(2) (i) When the daily noise exposure is composed of two or more periods of noise exposure of different levels, their combined effect should be considered, rather than the individual effect of each. Exposure to different levels for various periods of time shall be computed according to the formula set forth in paragraph (d)(2)(ii) of this section.

(ii) $F(e) = (T(1)\text{divided by } L(1)) + (T(2)\text{divided by } L(2)) + \ldots + (T(n)$ divided by $L(n))$ where:

$F(e)$ = The equivalent noise exposure factor.

T = The period of noise exposure at any essentially constant level.

L = The duration of the permissible noise exposure at the constant level (from Table D-2).

If the value of $F(e)$ exceeds unity (1) the exposure exceeds permissible levels.

(iii) A sample computation showing an application of the formula in paragraph (d)(2)(ii) of this section is as follows. An employee is exposed at these levels for these periods:

110 db for ¼ hour.

100 db for ½ hour.

90 db for 1½ hours.

$F(e)$ = (¼ divided by ½) + (½ divided by 2) + (1½ divided by 8)

$F(e) = 0.500 + 0.25 + 0.188$

$F(e) = 0.938$

Since the value of $F(e)$ does not exceed unity, the exposure is within permissible limits.

(e) Exposure to impulsive or impact noise should not exceed 140 dB peak sound pressure level.

1926.55—Gases, Vapors, Fumes, Dusts, and Mists

(a) Exposure of employees to inhalation, ingestion, skin absorption, or contact with any material or substance at a concentration above those specified in the "Threshold Limit Values of Airborne Contaminants for 1970" of the American Conference of Governmental Industrial Hygienists, shall be avoided.

(b) To achieve compliance with paragraph (a) of this section, administrative or engineering controls must first be implemented whenever feasible. When such controls are not feasible to achieve full compliance, protective equipment or other protective measures shall be used to keep the exposure of employees to air contaminants within the limits prescribed in this section. Any equipment and technical measures used for this purpose must first be approved for each particular use by a competent industrial hygienist or other technically qualified person. Whenever respirators are used, their use shall comply with 1926.103.

(c) Paragraphs (a) and (b) of this section do not apply to the exposure of employees to airborne asbestos, tremolite, anthophyllite, or actinolite dust. Whenever any employee is exposed to airborne asbestos, tremolite, anthophyllite, or actinolite dust, the requirements of 1910.1101 or 1926.58 of this title shall apply.

(d) Paragraphs (a) and (b) of this section do not apply to the exposure of employees to formaldehyde. Whenever any employee is exposed to formaldehyde, the requirements of 1910.1048 of this title shall apply.

1926.56—Illumination

(a) General. Construction areas, ramps, runways, corridors, offices, shops, and storage areas shall be lighted to not less than the minimum illumination intensities listed in Table D-3 while any work is in progress:

Table D-3 Minimum Illumination Intensities In Foot-Candles

Foot-Candles	Area of Operation
5	General construction area lighting.
3	General construction areas, concrete placement, excavation and waste areas, access ways, active storage areas, loading platforms, refueling, and field maintenance areas.
5	Indoors: warehouses, corridors, hallways, and exitways.
5	Tunnels, shafts, and general underground work areas (Exception: minimum of 10 foot-candles is required at tunnel and shaft heading during drilling, mucking, and scaling. Bureau of Mines approved cap lights shall be acceptable for use in the tunnel heading).
10	General construction plant and shops (e.g., batch plants, screening plants, mechanical and electrical equipment rooms, carpenter shops, rigging lofts and active store rooms, mess halls, and indoor toilets and workrooms).
30	First aid stations, infirmaries, and offices.

(b) Other areas. For areas or operations not covered above, refer to the American National Standard A11.1–1965, R1970, Practice for Industrial Lighting, for recommended values of illumination.

1926.59—Hazard Communication

Note: The requirements applicable to construction work under this section are identical to those set forth at 1910.1200 of this chapter.

1926 SUBPART E—PERSONAL PROTECTIVE AND LIFE SAVING EQUIPMENT

The requirement to use proper PPE is discussed throughout the book. Specifically, Chapter 9 illustrates how eye protection was particularly needed during Fire Station 39's interior construction due to the use of pneumatic tools.

1926.95—Criteria for Personal Protective Equipment

(a) Protective equipment, including personal protective equipment for eyes, face, head, and extremities, protective clothing, respiratory devices, and protective shields and barriers, shall be provided, used, and maintained in a sanitary and reliable condition wherever it is necessary by reason of hazards of processes or environment, chemical hazards, radiological hazards, or mechanical irritants encountered in a manner capable of causing injury or impairment in the function of any part of the body through absorption, inhalation or physical contact.

(b) Where employees provide their own protective equipment, the employer shall be responsible to assure its adequacy, including proper maintenance, and sanitation of such equipment.

(c) All personal protective equipment shall be of safe design and construction for the work to be performed.

(d) Payment for protective equipment.

(1) Except as provided by paragraphs (d)(2) through (d)(6) of this section, the protective equipment, including personal protective equipment (PPE), used to comply with this part, shall be provided by the employer at no cost to employees.

(2) The employer is not required to pay for non-specialty safety-toe protective footwear (including steel-toe shoes or steel-toe boots) and non-specialty prescription safety eyewear, provided that the employer permits such items to be worn off the job-site.

(3) When the employer provides metatarsal guards and allows the employee, at his or her request, to use shoes or boots with built-in metatarsal protection, the employer is not required to reimburse the employee for the shoes or boots.

(4) The employer is not required to pay for:
   (i) Everyday clothing, such as long-sleeve shirts, long pants, street shoes, and normal work boots; or
   (ii) Ordinary clothing, skin creams, or other items, used solely for protection from weather, such as winter coats, jackets, gloves, parkas, rubber boots, hats, raincoats, ordinary sunglasses, and sunscreen.

(5) The employer must pay for replacement PPE, except when the employee has lost or intentionally damaged the PPE.

(6) Where an employee provides adequate protective equipment he or she owns pursuant to paragraph (b) of this section, the employer may allow the employee to use it and is not required to reimburse the employee for that equipment. The employer shall not require an employee to provide or pay for his or her own PPE, unless the PPE is excepted by paragraphs (d)(2) through (d)(5) of this section.

1926.96—Occupational Foot Protection

Safety-toe footwear for employees shall meet the requirements and specifications in American National Standard for Men's Safety-Toe Footwear, Z41.1–1967.

1926.100—Head Protection

(a) Employees working in areas where there is a possible danger of head injury from impact, or from falling or flying objects, or from electrical shock and burns, shall be protected by protective helmets.

(b) Helmets for the protection of employees against impact and penetration of falling and flying objects shall meet the specifications contained in American National Standards Institute, Z89.1–1969, Safety Requirements for Industrial Head Protection.
   (1) The employer must provide each employee with head protection that meets the specifications.
   (2) The employer must ensure that the head protection provided for each employee exposed to high-voltage electric shock and burns also meets the specifications contained in Section 9.7 ("Electrical Insulation") of any of the consensus standards identified in paragraph (b)(1) of this section.
   (3) OSHA will deem any head protection device that the employer demonstrates is at least as effective as a head protection device constructed in accordance with one of the consensus standards identified in paragraph (b)(1) of this section to be in compliance with the requirements of this section.

(c) Helmets for the head protection of employees exposed to high voltage electrical shock and burns shall meet the specifications contained in American National Standards Institute, Z89.2–1971.

1926.101—Hearing Protection

(a) Wherever it is not feasible to reduce the noise levels or duration of exposures to those specified in Table D-2, Permissible Noise Exposures, in 1926.52, ear protective devices shall be provided and used.
(b) Ear protective devices inserted in the ear shall be fitted or determined individually by competent persons.
(c) Plain cotton is not an acceptable protective device.

1926.102—Eye and Face Protection

(a) General.
(1) Employees shall be provided with eye and face protection equipment when machines or operations present potential eye or face injury from physical, chemical, or radiation agents.
(2) Eye and face protection equipment required by this Part shall meet the requirements specified in American National Standards Institute, Z87.1–1968, Practice for Occupational and Educational Eye and Face Protection.
(3) Employees whose vision requires the use of corrective lenses in spectacles, when required by this regulation to wear eye protection, shall be protected by goggles or spectacles of one of the following types:
(i) Spectacles whose protective lenses provide optical correction;
(ii) Goggles that can be worn over corrective spectacles without disturbing the adjustment of the spectacles;
(iii) Goggles that incorporate corrective lenses mounted behind the protective lenses.
(4) Face and eye protection equipment shall be kept clean and in good repair. The use of this type [of] equipment with structural or optical defects shall be prohibited.
(5) Table E-1 shall be used as a guide in the selection of face and eye protection for the hazards and operations noted.
(6) Protectors shall meet the following minimum requirements:
(i) They shall provide adequate protection against the particular hazards for which they are designed.
(ii) They shall be reasonably comfortable when worn under the designated conditions.
(iii) They shall fit snugly and shall not unduly interfere with the movements of the wearer.
(iv) They shall be durable.
(v) They shall be capable of being disinfected.
(vi) They shall be easily cleanable.
(7) Every protector shall be distinctly marked to facilitate identification only of the manufacturer.
(8) When limitations or precautions are indicated by the manufacturer, they shall be transmitted to the user and care taken to see that such limitations and precautions are strictly observed.

Table E-1 Eye and Face Protector Selection Guide

1	Goggles	Flexible fitting	Regular ventilation
2	Goggles	Flexible fitting	Hooded ventilation
3	Goggles	Cushioned fitting	Rigid body
4	Spectacles	Metal frame	With sideshields (1)
5	Spectacles	Plastic frame	With sideshields (1)
6	Spectacles	Metal-plastic frame	with sideshields (1)
7	Welding goggles	Eyecup type	Tinted lenses (2)
7A	Chipping goggles	Eyecup type	Clear safety lenses
8	Welding goggles	Coverspec type (3)	Tinted lenses (2)
8A	Chipping goggles	Coverspec type (3)	Clear safety lenses
9	Welding goggles	Coverspec type (3)	Tinted plate lens (2)
10	Face shield		(Available with plastic or mesh window)
11	Welding helmets		(2)

Footnote (1) Non-side shield spectacles are available for limited hazard use requiring only frontal protection.
Footnote (2) See Table E-2, in paragraph (b) of this section, Filter Lens Shade Numbers for Protection Against Radiant Energy.
Footnote (3) Coverspec goggles can accommodate spectacles.

Table E-2 Filter Lens Shade Numbers for Protection Against Radiant Energy

Welding Operation	**Shade Number**
Shielded metal-arc welding 1/16-, 3/32-, 1/8-, 5/32-inch diameter electrodes	10
Gas-shielded arc welding (nonferrous) 1/16-, 3/32-, 1/8-, 5/32-inch diameter electrodes	11
Gas-shielded arc welding (ferrous) 1/16-, 3/32-, 1/8-, 5/32-inch diameter electrodes	12
Shielded metal-arc welding 3/16-, 7/32-, 1/4-inch diameter electrodes	12
5/16-, 3/8-inch diameter electrodes	14
Atomic hydrogen welding	10–14
Carbon-arc welding	14
Soldering	2
Torch brazing	3 or 4
Light cutting, up to 1 inch	3 or 4
Medium cutting, 1 inch to 6 inches	4 or 5
Heavy cutting, over 6 inches	5 or 6
Gas welding (light), up to 1/8-inch	4 or 5
Gas welding (medium), 1/8-inch to 1/2-inch	5 or 6
Gas welding (heavy), over 1/2-inch	6 or 8

(b) Protection against radiant energy-
   (1) Selection of shade numbers for welding filter. Table E-2 shall be used as a guide for the selection of the proper shade numbers of filter lenses or plates used in welding. Shades more dense than those listed may be used to suit the individual's needs.

1926.103—Respiratory Protection

At the Fire Station 39 project, workers excavating deeper than four feet for the underground cistern and some foundation footings might encounter hazardous atmosphere and would need to use NIOSH approved respirators.

Note: The requirements applicable to construction work under this section are identical to those set forth at 29 CFR 1910.134 of this chapter.

1926.104—Safety Belts, Lifelines, and Lanyards

(a) Lifelines, safety belts, and lanyards shall be used only for employee safeguarding. Any lifeline, safety belt, or lanyard actually subjected to in-service loading, as distinguished from static load testing, shall be immediately removed from service and shall not be used again for employee safeguarding.

(b) Lifelines shall be secured above the point of operation to an anchorage or structural member capable of supporting a minimum dead weight of 5,400 pounds.

(c) Lifelines used on rock-scaling operations, or in areas where the lifeline may be subjected to cutting or abrasion, shall be a minimum of ⅞-inch wire core manila rope. For all other lifeline applications, a minimum of ¾-inch manila or equivalent, with a minimum breaking strength of 5,400 pounds, shall be used.

(d) Safety belt lanyard shall be a minimum of ½-inch nylon, or equivalent, with a maximum length to provide for a fall of no greater than 6 feet. The rope shall have a nominal breaking strength of 5,400 pounds.

(e) All safety belt and lanyard hardware shall be drop forged or pressed steel, cadmium plated in accordance with type 1, Class B plating specified in Federal Specification QQ-P-416. Surface shall be smooth and free of sharp edges.

(f) All safety belt and lanyard hardware, except rivets, shall be capable of withstanding a tensile loading of 4,000 pounds without cracking, breaking, or taking a permanent deformation.

1926.105—Safety Nets

(a) Safety nets shall be provided when workplaces are more than 25 feet above the ground or water surface, or other surfaces where the use of ladders, scaffolds, catch platforms, temporary floors, safety lines, or safety belts is impractical.

(b) Where safety net protection is required by this part, operations shall not be undertaken until the net is in place and has been tested.

(c) (1) Nets shall extend 8 feet beyond the edge of the work surface where employees are exposed and shall be installed as close under the work surface as practical but in no case more than 25 feet below such work surface. Nets shall be hung with sufficient clearance to prevent user's contact with the surfaces or structures below. Such clearances shall be determined by impact load testing.

(2) It is intended that only one level of nets be required for bridge construction.

(d) The mesh size of nets shall not exceed 6 inches by 6 inches. All new nets shall meet accepted performance standards of 17,500 foot-pounds minimum impact resistance as determined and certified by the manufacturers, and shall bear a label of proof test. Edge ropes shall provide a minimum breaking strength of 5,000 pounds.

(e) Forged steel safety hooks or shackles shall be used to fasten the net to its supports.

(f) Connections between net panels shall develop the full strength of the net.

1926 SUBPART H—MATERIALS HANDLING, STORAGE, USE, AND DISPOSAL

1926.250—General Requirements for Storage

Masonry blocks and structural steel had to be stored on site at the Fire Station 39 project. Chapter 7 discusses the material storage requirement for masonry construction and steel erection.

(a) General.

(1) All materials stored in tiers shall be stacked, racked, blocked, interlocked, or otherwise secured to prevent sliding, falling or collapse.

(2) Maximum safe load limits of floors within buildings and structures, in pounds per square foot, shall be conspicuously posted in all storage areas, except for floor or slab on grade. Maximum safe loads shall not be exceeded.

(3) Aisles and passageways shall be kept clear to provide for the free and safe movement of material handling equipment or employees. Such areas shall be kept in good repair.

(4) When a difference in road or working levels exist, means such as ramps, blocking, or grading shall be used to ensure the safe movement of vehicles between the two levels.

(b) Material storage.

(1) Material stored inside buildings under construction shall not be placed within 6 feet of any hoistway or inside floor openings, nor within 10 feet of an exterior wall which does not extend above the top of the material stored.

(2) Each employee required to work on stored material in silos, hoppers, tanks, and similar storage areas shall be equipped with personal fall arrest equipment meeting the requirements of Subpart M of this part.

(3) Noncompatible materials shall be segregated in storage.

(4) Bagged materials shall be stacked by stepping back the layers and cross-keying the bags at least every 10 bags high.

(5) Materials shall not be stored on scaffolds or runways in excess of supplies needed for immediate operations.

(6) Brick stacks shall not be more than 7 feet in height. When a loose brick stack reaches a height of 4 feet, it shall be tapered back 2 inches in every foot of height above the 4-foot level.

(7) When masonry blocks are stacked higher than 6 feet, the stack shall be tapered back one-half block per tier above the 6-foot level.

(8) Lumber:

(i) Used lumber shall have all nails withdrawn before stacking.

(ii) Lumber shall be stacked on level and solidly supported sills.

(iii) Lumber shall be so stacked as to be stable and self-supporting.

(iv) Lumber piles shall not exceed 20 feet in height provided that lumber to be handled manually shall not be stacked more than 16 feet high.

(9) Structural steel, poles, pipe, bar stock, and other cylindrical materials, unless racked, shall be stacked and blocked so as to prevent spreading or tilting.

(c) Storage areas shall be kept free from accumulation of materials that constitute hazards from tripping, fire, explosion, or pest harborage. Vegetation control will be exercised when necessary.

1926.251—Rigging Equipment for Material Handling

Fire Station 39 is a structural steel building and required the use of a mobile crane to rig the structural steel members into their designated locations.

(a) General.

(1) Rigging equipment for material handling shall be inspected prior to use on each shift and as necessary during its use to ensure that it is safe. Defective rigging equipment shall be removed from service.

(2) Employers must ensure that rigging equipment:

(i) Has permanently affixed and legible identification markings as prescribed by the manufacturer that indicate the recommended safe working load;

(ii) Not be loaded in excess of its recommended safe working load as prescribed on the identification markings by the manufacturer; and

(iii) Not be used without affixed, legible identification markings, required by paragraph (a)(2)(i) of this section.

(3) Rigging equipment, when not in use, shall be removed from the immediate work area so as not to present a hazard to employees.

(4) Special custom design grabs, hooks, clamps, or other lifting accessories, for such units as modular panels, prefabricated structures and similar materials, shall be marked to indicate the safe working loads and shall be proof-tested prior to use to 125 percent of their rated load.

(5) This section applies to slings used in conjunction with other material handling equipment for the movement of material by hoisting, in employments covered by this part. The types of slings covered are those made from alloy steel chain, wire rope, metal mesh, natural or synthetic fiber rope (conventional three strand construction), and synthetic web (nylon, polyester, and polypropylene).

(6) Each day before being used, the sling and all fastenings and attachments shall be inspected for damage or defects by a competent person designated by the employer. Additional inspections shall be performed during sling use, where service conditions warrant. Damaged or defective slings shall be immediately removed from service.

(b) Alloy steel chains.

(1) Welded alloy steel chain slings shall have permanently affixed durable identification stating size, grade, rated capacity, and sling manufacturer.

(2) Hooks, rings, oblong links, pear-shaped links, welded or mechanical coupling links, or other attachments, when used with alloy steel chains, shall have a rated capacity at least equal to that of the chain.

(3) Job or shop hooks and links, or makeshift fasteners, formed from bolts, rods, etc., or other such attachments, shall not be used.

(4) Employers must not use alloy steel-chain slings with loads in excess of the rated capacities (i.e., working load limits) indicated on the sling by permanently affixed and legible identification markings prescribed by the manufacturer.

(5) Whenever wear at any point of any chain link exceeds that shown in Table H–1, the assembly shall be removed from service.

(6) "Inspections."

(i) In addition to the inspection required by other paragraphs of this section, a thorough periodic inspection of alloy steel chain slings in use shall be made on a regular basis, to be determined on the basis of (A) frequency of sling use; (B) severity of service conditions; (C) nature of lifts being made; and (D) experience gained on the service life of slings used in similar circumstances. Such inspections shall in no event be at intervals greater than once every 12 months.

(ii) The employer shall make and maintain a record of the most recent month in which each alloy steel chain sling was thoroughly inspected, and shall make such record available for examination.

(c) Wire rope.

(1) Employers must not use improved plow-steel wire rope and wire-rope slings with loads in excess of the rated capacities (i.e., working load limits) indicated on the sling by permanently affixed and legible identification markings prescribed by the manufacturer.

(2) Protruding ends of strands in splices on slings and bridles shall be covered or blunted.

(3) Wire rope shall not be secured by knots, except on haul back lines on scrapers.

(4) The following limitations shall apply to the use of wire rope:

(i) An eye splice made in any wire rope shall have not less than three full tucks. However, this requirement shall not operate to preclude the use of another form of splice or connection which can be shown to be as efficient and which is not otherwise prohibited.

(ii) Except for eye splices in the ends of wires and for endless rope slings, each wire rope used in hoisting or lowering, or in pulling loads, shall consist of one continuous piece without knot or splice.

(iii) Eyes in wire rope bridles, slings, or bull wires shall not be formed by wire rope clips or knots.

(iv) Wire rope shall not be used if, in any length of eight diameters, the total number of visible broken wires exceeds 10 percent of the total number of wires, or if the rope shows other signs of excessive wear, corrosion, or defect.

(5) When U-bolt wire rope clips are used to form eyes, Table H–2 shall be used to determine the number and spacing of clips. When used for eye splices, the U-bolt shall be applied so that the "U" section is in contact with the dead end of the rope.

(6) Slings shall not be shortened with knots or bolts or other makeshift devices.

(7) Sling legs shall not be kinked.

(8) Slings used in a basket hitch shall have the loads balanced to prevent slippage.

(9) Slings shall be padded or protected from the sharp edges of their loads.

(10) Hands or fingers shall not be placed between the sling and its load while the sling is being tightened around the load.

(11) Shock loading is prohibited.

(12) A sling shall not be pulled from under a load when the load is resting on the sling.

(13) "Minimum sling lengths."

(i) Cable laid and 6 X 19 and 6 X 37 slings shall have minimum clear length of wire rope 10 times the component rope diameter between splices, sleeves or end fittings.

(ii) Braided slings shall have a minimum clear length of wire rope 40 times the component rope diameter between the loops or end fittings.

(iii) Cable laid grommets, strand laid grommets and endless slings shall have a minimum circumferential length of 96 times their body diameter.

(14) Fiber core wire rope slings of all grades shall be permanently removed from service if they are exposed to temperatures in excess of 200 deg.

F (93.33 deg. C). When nonfiber core wire rope slings of any grade are used at temperatures above 400 deg. F (204.44 deg. C) or below minus 60 deg. F (15.55 deg. C), recommendations of the sling manufacturer regarding use at that temperature shall be followed.

(15) "End attachments."

(i) Welding of end attachments, except covers to thimbles, shall be performed prior to the assembly of the sling.

(ii) All welded end attachments shall not be used unless proof tested by the manufacturer or equivalent entity at twice their rated capacity prior to initial use. The employer shall retain a certificate of proof test, and make it available for examination.

(16) Wire rope slings shall have permanently affixed, legible identification markings stating size, rated capacity for the type(s) of hitch(es) used and the angle upon which it is based, and the number of legs if more than one.

(d) Natural rope, and synthetic fiber.

(1) Employers must not use natural- and synthetic-fiber rope slings with loads in excess of the rated capacities (i.e., working load limits) indicated on the sling by permanently affixed and legible identification markings prescribed by the manufacturer.

(2) All splices in rope slings provided by the employer shall be made in accordance with fiber rope manufacturers recommendations.

(i) In manila rope, eye splices shall contain at least three full tucks, and short splices shall contain at least six full tucks (three on each side of the center line of the splice).

(ii) In layed synthetic fiber rope, eye splices shall contain at least four full tucks, and short splices shall contain at least eight full tucks (four on each side of the center line of the splice).

(iii) Strand end tails shall not be trimmed short (flush with the surface of the rope) immediately adjacent to the full tucks. This precaution applies to both eye and short splices and all types of fiber rope. For fiber ropes under 1-inch diameter, the tails shall project at least six rope diameters beyond the last full tuck. For fiber ropes 1-inch diameter and larger, the tails shall project at least 6 inches beyond the last full tuck. In applications where the projecting tails may be objectionable, the tails shall be tapered and spliced into the body of the rope using at least two additional tucks (which will require a tail length of approximately six rope diameters beyond the last full tuck).

(iv) For all eye splices, the eye shall be sufficiently large to provide an included angle of not greater than 60 deg. at the splice when the eye is placed over the load or support.

(v) Knots shall not be used in lieu of splices.

(3) Natural and synthetic fiber rope slings, except for wet frozen slings, may be used in a temperature range from minus 20 deg. F (-28.88 deg. C) to plus 180 deg. F (82.2 deg. C) without decreasing the working load

limit. For operations outside this temperature range and for wet frozen slings, the sling manufacturer's recommendations shall be followed.

(4) Spliced fiber rope slings shall not be used unless they have been spliced in accordance with the following minimum requirements and in accordance with any additional recommendations of the manufacturer:

(i) In manila rope, eye splices shall consist of at least three full tucks, and short splices shall consist of at least six full tucks, three on each side of the splice center line.

(ii) In synthetic fiber rope, eye splices shall consist of at least four full tucks, and short splices shall consist of at least eight full tucks, four on each side of the center line.

(iii) Strand end tails shall not be trimmed flush with the surface of the rope immediately adjacent to the full tucks. This applies to all types of fiber rope and both eye and short splices. For fiber rope under 1 inch (2.54 cm) in diameter, the tail shall project at least six rope diameters beyond the last full tuck. For fiber rope 1 inch (2.54 cm) in diameter and larger, the tail shall project at least 6 inches (15.24 cm) beyond the last full tuck. Where a projecting tail interferes with the use of the sling, the tail shall be tapered and spliced into the body of the rope using at least two additional tucks (which will require a tail length of approximately six rope diameters beyond the last full tuck).

(iv) Fiber rope slings shall have a minimum clear length of rope between eye splices equal to 10 times the rope diameter.

(v) Knots shall not be used in lieu of splices.

(vi) Clamps not designed specifically for fiber ropes shall not be used for splicing.

(vii) For all eye splices, the eye shall be of such size to provide an included angle of not greater than 60 degrees at the splice when the eye is placed over the load or support.

(5) Fiber rope slings shall not be used if end attachments in contact with the rope have sharp edges or projections.

(6) Natural and synthetic fiber rope slings shall be immediately removed from service if any of the following conditions are present:

(i) Abnormal wear.

(ii) Powdered fiber between strands.

(iii) Broken or cut fibers.

(iv) Variations in the size or roundness of strands.

(v) Discoloration or rotting.

(vi) Distortion of hardware in the sling.

(7) Employers must use natural- and synthetic-fiber rope slings that have permanently affixed and legible identification markings that state the rated capacity for the type(s) of hitch(es) used and the angle upon which it is based, type of fiber material, and the number of legs if more than one.

(e) Synthetic webbing (nylon, polyester, and polypropylene).

(1) The employer shall have each synthetic web sling marked or coded to show:

(i) Name or trademark of manufacturer.

(ii) Rated capacities for the type of hitch.

(iii) Type of material.

(2) Rated capacity shall not be exceeded.

(3) Synthetic webbing shall be of uniform thickness and width and selvage edges shall not be split from the webbing's width.

(4) Fittings shall be:

(i) Of a minimum breaking strength equal to that of the sling; and

(ii) Free of all sharp edges that could in any way damage the webbing.

(5) Stitching shall be the only method used to attach end fittings to webbing and to form eyes. The thread shall be in an even pattern and contain a sufficient number of stitches to develop the full breaking strength of the sling.

(6) When synthetic web slings are used, the following precautions shall be taken:

(i) Nylon web slings shall not be used where fumes, vapors, sprays, mists or liquids of acids or phenolics are present.

(ii) Polyester and polypropylene web slings shall not be used where fumes, vapors, sprays, mists or liquids of caustics are present.

(iii) Web slings with aluminum fittings shall not be used where fumes, vapors, sprays, mists or liquids of caustics are present.

(7) Synthetic web slings of polyester and nylon shall not be used at temperatures in excess of 180 deg. F (82.2 deg. C). Polypropylene web slings shall not be used at temperatures in excess of 200 deg. F (93.33 deg. C).

(8) Synthetic web slings shall be immediately removed from service if any of the following conditions are present:

(i) Acid or caustic burns;

(ii) Melting or charring of any part of the sling surface;

(iii) Snags, punctures, tears or cuts;

(iv) Broken or worn stitches; or

(v) Distortion of fittings.

(f) Shackles and hooks.

(1) Employers must not use shackles with loads in excess of the rated capacities (i.e., working load limits) indicated on the shackle by permanently affixed and legible identification markings prescribed by the manufacturer.

(2) The manufacturer's recommendations shall be followed in determining the safe working loads of the various sizes and types of specific and identifiable hooks. All hooks for which no applicable manufacturer's recommendations are available shall be tested to twice the intended safe working load before they are initially put into use. The employer shall maintain a record of the dates and results of such tests.

1926.252—Disposal of Waste Materials

(a) Whenever materials are dropped more than 20 feet to any point lying outside the exterior walls of the building, an enclosed chute of wood, or equivalent material, shall be used. For the purpose of this paragraph, an enclosed chute is a slide, closed in on all sides, through which material is moved from a high place to a lower one.

(b) When debris is dropped through holes in the floor without the use of chutes, the area onto which the material is dropped shall be completely enclosed with barricades not less than 42 inches high and not less than 6 feet back from the projected edge of the opening above. Signs warning of the hazard of falling materials shall be posted at each level. Removal shall not be permitted in this lower area until debris handling ceases above.

(c) All scrap lumber, waste material, and rubbish shall be removed from the immediate work area as the work progresses.

(d) Disposal of waste material or debris by burning shall comply with local fire regulations.

(e) All solvent waste, oily rags, and flammable liquids shall be kept in fire resistant covered containers until removed from worksite.

1926 SUBPART L—SCAFFOLDS

1926.450—Scope, Application and Definitions Applicable To This Subpart

(a) Scope and application. This subpart applies to all scaffolds used in workplaces covered by this part. It does not apply to crane or derrick suspended personnel platforms. The criteria for aerial lifts are set out exclusively in § 1926.453.

1926.451—General Requirements

Chapter 7 discusses how scaffolding should be properly set up with necessary fall protection at the Fire Station 39 project during masonry wall construction. The standards below state the general requirements for proper scaffold setup and the need to have fall protection. Detail fall protection standards can be found in Subpart M.

This section does not apply to aerial lifts, the criteria for which are set out exclusively in 1926.453.

(a) (1) Except as provided in paragraphs (a)(2), (a)(3), (a)(4), (a)(5) and (g) of this section, each scaffold and scaffold component shall be capable of supporting, without failure, its own weight and at least 4 times the maximum intended load applied or transmitted to it.

(2) Direct connections to roofs and floors, and counterweights used to balance adjustable suspension scaffolds, shall be capable of resisting at least 4 times the tipping moment imposed by the scaffold operating at the

rated load of the hoist, or 1.5 (minimum) times the tipping moment imposed by the scaffold operating at the stall load of the hoist, whichever is greater.

(3) Each suspension rope, including connecting hardware, used on non-adjustable suspension scaffolds shall be capable of supporting, without failure, at least 6 times the maximum intended load applied or transmitted to that rope.

(4) Each suspension rope, including connecting hardware, used on adjustable suspension scaffolds shall be capable of supporting, without failure, at least 6 times the maximum intended load applied or transmitted to that rope with the scaffold operating at either the rated load of the hoist, or 2 (minimum) times the stall load of the hoist, whichever is greater.

(5) The stall load of any scaffold hoist shall not exceed 3 times its rated load.

(6) Scaffolds shall be designed by a qualified person and shall be constructed and loaded in accordance with that design. Non-mandatory Appendix A to this subpart contains examples of criteria that will enable an employer to comply with paragraph (a) of this section.

(b) (1) Each platform on all working levels of scaffolds shall be fully planked or decked between the front uprights and the guardrail supports as follows:

(i) Each platform unit (e.g., scaffold plank, fabricated plank, fabricated deck, or fabricated platform) shall be installed so that the space between adjacent units and the space between the platform and the uprights is no more than 1 inch (2.5 cm) wide, except where the employer can demonstrate that a wider space is necessary (for example, to fit around uprights when side brackets are used to extend the width of the platform).

(ii) Where the employer makes the demonstration provided for in paragraph (b)(1)(i) of this section, the platform shall be planked or decked as fully as possible and the remaining open space between the platform and the uprights shall not exceed 9½ inches (24.1 cm).

Exception to paragraph (b)(1): The requirement in paragraph (b)(1) to provide full planking or decking does not apply to platforms used solely as walkways or solely by employees performing scaffold erection or dismantling. In these situations, only the planking that the employer establishes is necessary to provide safe working conditions is required.

(2) Except as provided in paragraphs (b)(2)(i) and (b)(2)(ii) of this section, each scaffold platform and walkway shall be at least 18 inches (46 cm) wide.

(i) Each ladder jack scaffold, top plate bracket scaffold, roof bracket scaffold, and pump jack scaffold shall be at least 12 inches (30 cm) wide. There is no minimum width requirement for boatswains' chairs.

(ii) Where scaffolds must be used in areas that the employer can demonstrate are so narrow that platforms and walkways cannot be at least 18 inches (46 cm) wide, such platforms and walkways shall be as wide as feasible, and employees on those platforms and walkways shall be protected from fall hazards by the use of guardrails and/or personal fall arrest systems.

(3) Except as provided in paragraphs (b)(3)(i) and (ii) of this section, the front edge of all platforms shall not be more than 14 inches (36 cm) from the face of the work, unless guardrail systems are erected along the front edge and/or personal fall arrest systems are used in accordance with paragraph (g) of this section to protect employees from falling.

(i) The maximum distance from the face for outrigger scaffolds shall be 3 inches (8 cm);

(ii) The maximum distance from the face for plastering and lathing operations shall be 18 inches (46 cm).

(4) Each end of a platform, unless cleated or otherwise restrained by hooks or equivalent means, shall extend over the centerline of its support at least 6 inches (15 cm).

(5) (i) Each end of a platform 10 feet or less in length shall not extend over its support more than 12 inches (30 cm) unless the platform is designed and installed so that the cantilevered portion of the platform is able to support employees and/or materials without tipping, or has guardrails which block employee access to the cantilevered end.

(ii) Each platform greater than 10 feet in length shall not extend over its support more than 18 inches (46 cm), unless it is designed and installed so that the cantilevered portion of the platform is able to support employees without tipping, or has guardrails which block employee access to the cantilevered end.

(6) On scaffolds where scaffold planks are abutted to create a long platform, each abutted end shall rest on a separate support surface. This provision does not preclude the use of common support members, such as "T" sections, to support abutting planks, or hook on platforms designed to rest on common supports.

(7) On scaffolds where platforms are overlapped to create a long platform, the overlap shall occur only over supports, and shall not be less than 12 inches (30 cm) unless the platforms are nailed together or otherwise restrained to prevent movement.

(8) At all points of a scaffold where the platform changes direction, such as turning a corner, any platform that rests on a bearer at an angle other than a right angle shall be laid first, and platforms which rest at right angles over the same bearer shall be laid second, on top of the first platform.

(9) Wood platforms shall not be covered with opaque finishes, except that platform edges may be covered or marked for identification. Platforms may be coated periodically with wood preservatives, fire-retardant finishes, and slip-resistant finishes; however, the coating may not obscure the top or bottom wood surfaces.

(10) Scaffold components manufactured by different manufacturers shall not be intermixed unless the components fit together without force and the scaffold's structural integrity is maintained by the user. Scaffold components manufactured by different manufacturers shall not be modified in order to intermix them unless a competent person determines the resulting scaffold is structurally sound.

(11) Scaffold components made of dissimilar metals shall not be used together unless a competent person has determined that galvanic action will not reduce the strength of any component to a level below that required by paragraph (a)(1) of this section.

(c) (1) Supported scaffolds with a height to base width (including outrigger supports, if used) ratio of more than four to one (4:1) shall be restrained from tipping by guying, tying, bracing, or equivalent means, as follows:

(i) Guys, ties, and braces shall be installed at locations where horizontal members support both inner and outer legs.

(ii) Guys, ties, and braces shall be installed according to the scaffold manufacturer's recommendations or at the closest horizontal member to the 4:1 height and be repeated vertically at locations of horizontal members every 20 feet (6.1 m) or less thereafter for scaffolds 3 feet (0.91 m) wide or less, and every 26 feet (7.9 m) or less thereafter for scaffolds greater than 3 feet (0.91 m) wide. The top guy, tie or brace of completed scaffolds shall be placed no further than the 4:1 height from the top. Such guys, ties and braces shall be installed at each end of the scaffold and at horizontal intervals not to exceed 30 feet (9.1 m) (measured from one end [not both] towards the other).

(iii) Ties, guys, braces, or outriggers shall be used to prevent the tipping of supported scaffolds in all circumstances where an eccentric load, such as a cantilevered work platform, is applied or is transmitted to the scaffold.

(2) Supported scaffold poles, legs, posts, frames, and uprights shall bear on base plates and mud sills or other adequate firm foundation.

(i) Footings shall be level, sound, rigid, and capable of supporting the loaded scaffold without settling or displacement.

(ii) Unstable objects shall not be used to support scaffolds or platform units.

(iii) Unstable objects shall not be used as working platforms.

(iv) Front-end loaders and similar pieces of equipment shall not be used to support scaffold platforms unless they have been specifically designed by the manufacturer for such use.

(v) Fork-lifts shall not be used to support scaffold platforms unless the entire platform is attached to the fork and the fork-lift is not moved horizontally while the platform is occupied.

(3) Supported scaffold poles, legs, posts, frames, and uprights shall be plumb and braced to prevent swaying and displacement.

(d) "Criteria for suspension scaffolds."

(e) This paragraph applies to scaffold access for all employees. Access requirements for employees erecting or dismantling supported scaffolds are specifically addressed in paragraph (e)(9) of this section.

(1) When scaffold platforms are more than 2 feet (0.6 m) above or below a point of access, portable ladders, hook-on ladders, attachable ladders, stair towers (scaffold stairways/towers), stairway-type ladders (such as ladder stands), ramps, walkways, integral prefabricated scaffold access, or direct access from another scaffold, structure, personnel hoist, or similar surface shall be used. Crossbraces shall not be used as a means of access.

(2) Portable, hook-on, and attachable ladders;

(i) Portable, hook-on, and attachable ladders shall be positioned so as not to tip the scaffold;

(ii) Hook-on and attachable ladders shall be positioned so that their bottom rung is not more than 24 inches (61 cm) above the scaffold supporting level;

(iii) When hook-on and attachable ladders are used on a supported scaffold more than 35 feet (10.7 m) high, they shall have rest platforms at 35-foot (10.7 m) maximum vertical intervals;

(iv) Hook-on and attachable ladders shall be specifically designed for use with the type of scaffold used;

(v) Hook-on and attachable ladders shall have a minimum rung length of 11½ inches (29 cm); and

(vi) Hook-on and attachable ladders shall have uniformly spaced rungs with a maximum spacing between rungs of 16¾ inches.

(3) Stairway-type ladders shall:

(i) Be positioned such that their bottom step is not more than 24 inches (61 cm) above the scaffold supporting level;

(ii) Be provided with rest platforms at 12 foot (3.7 m) maximum vertical intervals;

(iii) Have a minimum step width of 16 inches (41 cm), except that mobile scaffold stairway-type ladders shall have a minimum step width of 11½ inches (30 cm); and

(iv) Have slip-resistant treads on all steps and landings.

(4) Stairtowers (scaffold stairway/towers) shall be positioned such that their bottom step is not more than 24 inches (61 cm.) above the scaffold supporting level.

(5) Ramps and walkways.

(i) Ramps and walkways 6 feet (1.8 m) or more above lower levels shall have guardrail systems which comply with subpart M of this part—Fall Protection.

(ii) No ramp or walkway shall be inclined more than a slope of one (1) vertical to three (3) horizontal (20 degrees above the horizontal).

(iii) If the slope of a ramp or a walkway is steeper than one (1) vertical in eight (8) horizontal, the ramp or walkway shall have cleats not more than fourteen (14) inches (35 cm) apart which are securely fastened to the planks to provide footing.

(6) Integral prefabricated scaffold access frames shall:
 (i) Be specifically designed and constructed for use as ladder rungs;
 (ii) Have a rung length of at least 8 inches (20 cm);
 (iii) Not be used as work platforms when rungs are less than 11½ inches in length, unless each affected employee uses fall protection, or a positioning device, which complies with 1926.502;
 (iv) Be uniformly spaced within each frame section;
 (v) Be provided with rest platforms at 35-foot (10.7 m) maximum vertical intervals on all supported scaffolds more than 35 feet (10.7 m) high; and
 (vi) Have a maximum spacing between rungs of 16¾ inches (43 cm). Non-uniform rung spacing caused by joining end frames together is allowed, provided the resulting spacing does not exceed 16¾ inches (43 cm).

(7) Steps and rungs of ladder and stairway type access shall line up vertically with each other between rest platforms.

(8) Direct access to or from another surface shall be used only when the scaffold is not more than 14 inches (36 cm) horizontally and not more than 24 inches (61 cm) vertically from the other surface.

(9) Effective September 2, 1997, access for employees erecting or dismantling supported scaffolds shall be in accordance with the following:
 (i) The employer shall provide safe means of access for each employee erecting or dismantling a scaffold where the provision of safe access is feasible and does not create a greater hazard. The employer shall have a competent person determine whether it is feasible or would pose a greater hazard to provide, and have employees use a safe means of access. This determination shall be based on site conditions and the type of scaffold being erected or dismantled.
 (ii) Hook-on or attachable ladders shall be installed as soon as scaffold erection has progressed to a point that permits safe installation and use.
 (iii) When erecting or dismantling tubular welded frame scaffolds, (end) frames, with horizontal members that are parallel, level and are not more than 22 inches apart vertically may be used as climbing devices for access, provided they are erected in a manner that creates a usable ladder and provides good hand hold and foot space.
 (iv) Cross braces on tubular welded frame scaffolds shall not be used as a means of access or egress.

(f) (1) Scaffolds and scaffold components shall not be loaded in excess of their maximum intended loads or rated capacities, whichever is less.

(2) The use of shore or lean-to scaffolds is prohibited.

(3) Scaffolds and scaffold components shall be inspected for visible defects by a competent person before each work shift, and after any occurrence which could affect a scaffold's structural integrity.

(4) Any part of a scaffold damaged or weakened such that its strength is less than that required by paragraph (a) of this section shall be immediately repaired or replaced, braced to meet those provisions, or removed from service until repaired.

(5) Scaffolds shall not be moved horizontally while employees are on them, unless they have been designed by a registered professional engineer specifically for such movement or, for mobile scaffolds, where the provisions of 1926.452(w) are followed.

(6) The clearance between scaffolds and power lines shall be as follows: Scaffolds shall not be erected, used, dismantled, altered, or moved such that they or any conductive material handled on them might come closer to exposed and energized power lines than as follows:

Exception to paragraph (f)(6): Scaffolds and materials may be closer to power lines than specified above where such clearance is necessary for performance of work, and only after the utility company, or electrical system operator, has been notified of the need to work closer and the utility company, or electrical system operator, has deenergized the lines, relocated the lines, or installed protective coverings to prevent accidental contact with the lines.

(7) Scaffolds shall be erected, moved, dismantled, or altered only under the supervision and direction of a competent person qualified in scaffold erection, moving, dismantling or alteration. Such activities shall be performed only by experienced and trained employees selected for such work by the competent person.

***Insulated Lines**

Voltage	Minimum Distance	Alternatives
Less than 300 volts	3 feet (0.9 m)	
300 volts to 50 kV	10 feet (3.1 m)	
More than 50 kV	10 feet (3.1 m) plus 0.4 inches (1.0 cm) for each 1 kV over 50 kV.	2 times the length of the line insulator, but never less than 10 feet (3.1 m).

***Uninsulated Lines**

Voltage	Minimum Distance	Alternatives
Less than 50 kV	10 feet (3.1 m).	
More than 50 kV	10 feet (3.1 m) plus 0.4 inches (1.0 cm) for each 1 kV over 50 kV.	2 times the length of the line insulator, but never less than 10 feet (3.1 m).

(8) Employees shall be prohibited from working on scaffolds covered with snow, ice, or other slippery material except as necessary for removal of such materials.

(9) Where swinging loads are being hoisted onto or near scaffolds such that the loads might contact the scaffold, tag lines or equivalent measures to control the loads shall be used.

(10) Suspension ropes supporting adjustable suspension scaffolds shall be of a diameter large enough to provide sufficient surface area for the functioning of brake and hoist mechanisms.

(11) Suspension ropes shall be shielded from heat-producing processes. When acids or other corrosive substances are used on a scaffold, the ropes shall be shielded, treated to protect against the corrosive substances, or shall be of a material that will not be damaged by the substance being used.

(12) Work on or from scaffolds is prohibited during storms or high winds unless a competent person has determined that it is safe for employees to be on the scaffold and those employees are protected by a personal fall arrest system or wind screens. Wind screens shall not be used unless the scaffold is secured against the anticipated wind forces imposed.

(13) Debris shall not be allowed to accumulate on platforms.

(14) Makeshift devices, such as but not limited to boxes and barrels, shall not be used on top of scaffold platforms to increase the working level height of employees.

(15) Ladders shall not be used on scaffolds to increase the working level height of employees, except on large area scaffolds where employers have satisfied the following criteria:

(i) When the ladder is placed against a structure which is not a part of the scaffold, the scaffold shall be secured against the sideways thrust exerted by the ladder;

(ii) The platform units shall be secured to the scaffold to prevent their movement;

(iii) The ladder legs shall be on the same platform or other means shall be provided to stabilize the ladder against unequal platform deflection; and

(iv) The ladder legs shall be secured to prevent them from slipping or being pushed off the platform.

(16) Platforms shall not deflect more than 1/60 of the span when loaded.

(17) To reduce the possibility of welding current arcing through the suspension wire rope when performing welding from suspended scaffolds, the following precautions shall be taken, as applicable:

(i) An insulated thimble shall be used to attach each suspension wire rope to its hanging support (such as cornice hook or outrigger). Excess suspension wire rope and any additional independent lines from grounding shall be insulated;

(ii) The suspension wire rope shall be covered with insulating material extending at least 4 feet (1.2 m) above the hoist. If there is a tail line below the hoist, it shall be insulated to prevent contact with the platform. The portion of the tail line that hangs free below the scaffold shall be guided or retained, or both, so that it does not become grounded;

(iii) Each hoist shall be covered with insulated protective covers;

(iv) In addition to a work lead attachment required by the welding process, a grounding conductor shall be connected from the scaffold to the structure. The size of this conductor shall be at least the size of the welding process work lead, and this conductor shall not be in series with the welding process or the work piece;

(v) If the scaffold grounding lead is disconnected at any time, the welding machine shall be shut off; and

(vi) An active welding rod or uninsulated welding lead shall not be allowed to contact the scaffold or its suspension system.

(g) (1) Each employee on a scaffold more than 10 feet (3.1 m) above a lower level shall be protected from falling to that lower level. Paragraphs (g)(1)(i) through (vii) of this section establish the types of fall protection to be provided to the employees on each type of scaffold. Paragraph (g)(2) of this section addresses fall protection for scaffold erectors and dismantlers.

Note to paragraph (g)(1): The fall protection requirements for employees installing suspension scaffold support systems on floors, roofs, and other elevated surfaces are set forth in subpart M of this part.

(1) (i) Each employee on a boatswains' chair, catenary scaffold, float scaffold, needle beam scaffold, or ladder jack scaffold shall be protected by a personal fall arrest system;

(ii) Each employee on a single-point or two-point adjustable suspension scaffold shall be protected by both a personal fall arrest system and guardrail system;

(iii) Each employee on a crawling board (chicken ladder) shall be protected by a personal fall arrest system, a guardrail system (with minimum 200 pound toprail capacity), or by a three-fourth inch (1.9 cm) diameter grabline or equivalent handhold securely fastened beside each crawling board;

(iv) Each employee on a self-contained adjustable scaffold shall be protected by a guardrail system (with minimum 200 pound toprail capacity) when the platform is supported by the frame structure, and by both a personal fall arrest system and a guardrail system (with minimum 200 pound toprail capacity) when the platform is supported by ropes;

(v) Each employee on a walkway located within a scaffold shall be protected by a guardrail system (with minimum 200 pound toprail capacity) installed within 9½ inches (24.1 cm) of and along at least one side of the walkway;

(vi) Each employee performing overhand bricklaying operations from a supported scaffold shall be protected from falling from all open sides and ends of the scaffold (except at the side next to the wall being laid) by the use of a personal fall arrest system or guardrail system (with minimum 200 pound toprail capacity);

(vii) For all scaffolds not otherwise specified in paragraphs (g)(1)(i) through (g)(1)(vi) of this section, each employee shall be protected by the use of personal fall arrest systems or guardrail systems meeting the requirements of paragraph (g)(4) of this section.

(2) The employer shall have a competent person determine the feasibility and safety of providing fall protection for employees erecting or dismantling supported scaffolds. Employers are required to provide fall protection for employees erecting or dismantling supported scaffolds where the installation and use of such protection is feasible and does not create a greater hazard.

(3) In addition to meeting the requirements of 1926.502(d), personal fall arrest systems used on scaffolds shall be attached by lanyard to a vertical lifeline, horizontal lifeline, or scaffold structural member. Vertical lifelines shall not be used when overhead components, such as overhead protection or additional platform levels, are part of a single-point or two-point adjustable suspension scaffold.

(i) When vertical lifelines are used, they shall be fastened to a fixed safe point of anchorage, shall be independent of the scaffold, and shall be protected from sharp edges and abrasion. Safe points of anchorage include structural members of buildings, but do not include standpipes, vents, other piping systems, electrical conduit, outrigger beams, or counterweights.

(ii) When horizontal lifelines are used, they shall be secured to two or more structural members of the scaffold, or they may be looped around both suspension and independent suspension lines (on scaffolds so equipped) above the hoist and brake attached to the end of the scaffold. Horizontal lifelines shall not be attached only to the suspension ropes.

(iii) When lanyards are connected to horizontal lifelines or structural members on a single-point or two-point adjustable suspension scaffold, the scaffold shall be equipped with additional independent support lines and automatic locking devices capable of stopping the fall of the scaffold in the event one or both of the suspension ropes fail. The independent support lines shall be equal in number and strength to the suspension ropes.

(iv) Vertical lifelines, independent support lines, and suspension ropes shall not be attached to each other, nor shall they be attached to or use the same point of anchorage, nor shall they be attached to the same point on the scaffold or personal fall arrest system.

(4) Guardrail systems installed to meet the requirements of this section shall comply with the following provisions (guardrail systems built in accordance with Appendix A to this subpart will be deemed to meet the requirements of paragraphs (g)(4)(vii), (viii), and (ix) of this section):

(i) Guardrail systems shall be installed along all open sides and ends of platforms. Guardrail systems shall be installed before the scaffold is released for use by employees other than erection/dismantling crews.

(ii) The top edge height of toprails or equivalent member on supported scaffolds manufactured or placed in service after January 1, 2000 shall be installed between 38 inches (0.97 m) and 45 inches (1.2 m) above the platform surface. The top edge height on supported scaffolds manufactured and placed in service before January 1, 2000, and on all suspended scaffolds where both a guardrail and a personal fall arrest system are required shall be between 36 inches (0.9 m) and 45 inches (1.2 m). When conditions warrant, the height of the top edge may exceed the 45-inch height, provided the guardrail system meets all other criteria of paragraph (g)(4).

(iii) When midrails, screens, mesh, intermediate vertical members, solid panels, or equivalent structural members are used, they shall be installed between the top edge of the guardrail system and the scaffold platform.

(iv) When midrails are used, they shall be installed at a height approximately midway between the top edge of the guardrail system and the platform surface.

(v) When screens and mesh are used, they shall extend from the top edge of the guardrail system to the scaffold platform, and along the entire opening between the supports.

(vi) When intermediate members (such as balusters or additional rails) are used, they shall not be more than 19 inches (48 cm) apart.

(vii) Each toprail or equivalent member of a guardrail system shall be capable of withstanding, without failure, a force applied in any downward or horizontal direction at any point along its top edge of at least 100 pounds (445 n) for guardrail systems installed on single-point adjustable suspension scaffolds or two-point adjustable suspension scaffolds, and at least 200 pounds (890 n) for guardrail systems installed on all other scaffolds.

(viii) When the loads specified in paragraph (g)(4)(vii) of this section are applied in a downward direction, the top edge shall not drop below the height above the platform surface that is prescribed in paragraph (g)(4)(ii) of this section.

(ix) Midrails, screens, mesh, intermediate vertical members, solid panels, and equivalent structural members of a guardrail system shall be capable of withstanding, without failure, a force applied in any downward or horizontal direction at any point along the midrail or other member of at least 75 pounds (333 n) for guardrail systems with a minimum 100 pound toprail capacity, and at least 150 pounds

(666 n) for guardrail systems with a minimum 200 pound toprail capacity.

(x) Suspension scaffold hoists and non-walk-through stirrups may be used as end guardrails, if the space between the hoist or stirrup and the side guardrail or structure does not allow passage of an employee to the end of the scaffold.

(xi) Guardrails shall be surfaced to prevent injury to an employee from punctures or lacerations, and to prevent snagging of clothing.

(xii) The ends of all rails shall not overhang the terminal posts except when such overhang does not constitute a projection hazard to employees.

(xiii) Steel or plastic banding shall not be used as a toprail or midrail.

(xiv) Manila or plastic (or other synthetic) rope being used for toprails or midrails shall be inspected by a competent person as frequently as necessary to ensure that it continues to meet the strength requirements of paragraph (g) of this section.

(xv) Crossbracing is acceptable in place of a midrail when the crossing point of two braces is between 20 inches (0.5 m) and 30 inches (0.8 m) above the work platform or as a toprail when the crossing point of two braces is between 38 inches (0.97 m) and 48 inches (1.3 m) above the work platform. The end points at each upright shall be no more than 48 inches (1.3 m) apart.

(h) (1) In addition to wearing hardhats each employee on a scaffold shall be provided with additional protection from falling hand tools, debris, and other small objects through the installation of toeboards, screens, or guardrail systems, or through the erection of debris nets, catch platforms, or canopy structures that contain or deflect the falling objects. When the falling objects are too large, heavy or massive to be contained or deflected by any of the above-listed measures, the employer shall place such potential falling objects away from the edge of the surface from which they could fall and shall secure those materials as necessary to prevent their falling.

(2) Where there is a danger of tools, materials, or equipment falling from a scaffold and striking employees below, the following provisions apply:

(i) The area below the scaffold to which objects can fall shall be barricaded, and employees shall not be permitted to enter the hazard area; or

(ii) A toeboard shall be erected along the edge of platforms more than 10 feet (3.1 m) above lower levels for a distance sufficient to protect employees below, except on float (ship) scaffolds where an edging of ¾ × 1½ inch (2 × 4 cm) wood or equivalent may be used in lieu of toeboards;

(iii) Where tools, materials, or equipment are piled to a height higher than the top edge of the toeboard, paneling or screening extending from the toeboard or platform to the top of the guardrail shall be erected for a distance sufficient to protect employees below; or

(iv) A guardrail system shall be installed with openings small enough to prevent passage of potential falling objects; or

(v) A canopy structure, debris net, or catch platform strong enough to withstand the impact forces of the potential falling objects shall be erected over the employees below.

(3) Canopies, when used for falling object protection, shall comply with the following criteria:

(i) Canopies shall be installed between the falling object hazard and the employees.

(ii) When canopies are used on suspension scaffolds for falling object protection, the scaffold shall be equipped with additional independent support lines equal in number to the number of points supported, and equivalent in strength to the strength of the suspension ropes.

(iii) Independent support lines and suspension ropes shall not be attached to the same points of anchorage.

(4) Where used, toeboards shall be:

(i) Capable of withstanding, without failure, a force of at least 50 pounds (222 N) applied in any downward or horizontal direction at any point along the toeboard (toeboards built in accordance with Appendix A to this subpart will be deemed to meet this requirement); and

(ii) At least three and one-half inches (9 cm) high from the top edge of the toeboard to the level of the walking/working surface. Toeboards shall be securely fastened in place at the outermost edge of the platform and have not more than ¼ inch (0.7 cm) clearance above the walking/working surface. Toeboards shall be solid or with openings not over one inch (2.5 cm) in the greatest dimension.

1926.452—Additional Requirements Applicable To Specific Types of Scaffolds

Tubular welded frame scaffolds and mobile scaffolds were employed at the Fire Station 39 project during masonry construction and interior construction, respectively. Chapter 7 discusses how the tubular welded frame scaffold was used on-site and Chapter 9 examines the application of the mobile scaffold.

In addition to the applicable requirements of 1926.451, the following requirements apply to the specific types of scaffolds indicated. Scaffolds not specifically addressed by 1926.452, such as but not limited to systems scaffolds, must meet the requirements of 1926.451.

(a) "Fabricated frame scaffolds" (tubular welded frame scaffolds).

(1) When moving platforms to the next level, the existing platform shall be left undisturbed until the new end frames have been set in place and braced prior to receiving the new platforms.

(2) Frames and panels shall be braced by cross, horizontal, or diagonal braces, or combination thereof, which secure vertical members together laterally. The cross braces shall be of such length as will automatically square and align vertical members so that the erected scaffold is always plumb, level, and square. All brace connections shall be secured.
(3) Frames and panels shall be joined together vertically by coupling or stacking pins or equivalent means.
(4) Where uplift can occur which would displace scaffold end frames or panels, the frames or panels shall be locked together vertically by pins or equivalent means.
(5) Brackets used to support cantilevered loads shall:
 (i) Be seated with side-brackets parallel to the frames and end-brackets at 90 degrees to the frames;
 (ii) Not be bent or twisted from these positions; and
 (iii) Be used only to support personnel, unless the scaffold has been designed for other loads by a qualified engineer and built to withstand the tipping forces caused by those other loads being placed on the bracket-supported section of the scaffold.
(6) Scaffolds over 125 feet (38.0 m) in height above their base plates shall be designed by a registered professional engineer, and shall be constructed and loaded in accordance with such design.

(b) "Plasterers', decorators', and large area scaffolds." Scaffolds shall be constructed in accordance with paragraph (a) of this section, as appropriate.

(c) "Two-point adjustable suspension scaffolds (swing stages)." The following requirements do not apply to two-point adjustable suspension scaffolds used as masons' or stonesetters' scaffolds. Such scaffolds are covered by paragraph (q) of this section.

(d) "Mobile scaffolds."
(1) Scaffolds shall be braced by cross, horizontal, or diagonal braces, or combination thereof, to prevent racking or collapse of the scaffold and to secure vertical members together laterally so as to automatically square and align the vertical members. Scaffolds shall be plumb, level, and squared. All brace connections shall be secured.
 (i) Scaffolds constructed of tube and coupler components shall also comply with the requirements of paragraph (b) of this section;
 (ii) Scaffolds constructed of fabricated frame components shall also comply with the requirements of paragraph (c) of this section.
(2) Scaffold casters and wheels shall be locked with positive wheel and/or wheel and swivel locks, or equivalent means, to prevent movement of the scaffold while the scaffold is used in a stationary manner.
(3) Manual force used to move the scaffold shall be applied as close to the base as practicable, but not more than 5 feet (1.5 m) above the supporting surface.
(4) Power systems used to propel mobile scaffolds shall be designed for such use. Forklifts, trucks, similar motor vehicles or add-on motors shall not be used to propel scaffolds unless the scaffold is designed for such propulsion systems.

(5) Scaffolds shall be stabilized to prevent tipping during movement.

(6) Employees shall not be allowed to ride on scaffolds unless the following conditions exist:

(i) The surface on which the scaffold is being moved is within 3 degrees of level, and free of pits, holes, and obstructions;

(ii) The height to base width ratio of the scaffold during movement is two to one or less;

(iii) Outrigger frames, when used, are installed on both sides of the scaffold;

(iv) When power systems are used, the propelling force is applied directly to the wheels, and does not produce a speed in excess of 1 foot per second (.3 mps); and

(v) No employee is on any part of the scaffold which extends outward beyond the wheels, casters, or other supports.

(7) Platforms shall not extend outward beyond the base supports of the scaffold unless outrigger frames or equivalent devices are used to ensure stability.

(8) Where leveling of the scaffold is necessary, screw jacks or equivalent means shall be used.

(9) Caster stems and wheel stems shall be pinned or otherwise secured in scaffold legs or adjustment screws.

(10) Before a scaffold is moved, each employee on the scaffold shall be made aware of the move.

(e) Stilts, when used, shall be used in accordance with the following requirements:

(1) An employee may wear stilts on a scaffold only if it is a large area scaffold.

(2) When an employee is using stilts on a large area scaffold where a guardrail system is used to provide fall protection, the guardrail system shall be increased in height by an amount equal to the height of the stilts being used by the employee.

(3) Surfaces on which stilts are used shall be flat and free of pits, holes and obstructions, such as debris, as well as other tripping and falling hazards.

(4) Stilts shall be properly maintained. Any alteration of the original equipment shall be approved by the manufacturer.

1926.453—Aerial Lifts

Chapter 8 discusses how an aerial lift was used on the Fire Station 39 project during exterior sheathing installation. This part of the OSHA standards considers aerial lift as a type of scaffold and describes how such a piece of equipment should be used with applicable fall protection requirement.

(a) (1) Unless otherwise provided in this section, aerial lifts acquired for use shall be designed and constructed in conformance with the applicable requirements of the American National Standards for "Vehicle Mounted Elevating and Rotating Work Platforms," ANSI A92.2–1969, including appendix. Aerial lifts include the following types of vehicle-mounted aerial devices used to elevate personnel to job-sites above ground:

(i) Extensible boom platforms;
(ii) Aerial ladders;
(iii) Articulating boom platforms;
(iv) Vertical towers; and
(v) A combination of any such devices. Aerial equipment may be made of metal, wood, fiberglass reinforced plastic (FRP), or other material; may be powered or manually operated; and are deemed to be aerial lifts whether or not they are capable of rotating about a substantially vertical axis.

(2) Aerial lifts may be "field modified" for uses other than those intended by the manufacturer provided the modification has been certified in writing by the manufacturer or by any other equivalent entity, such as a nationally recognized testing laboratory, to be in conformity with all applicable provisions of ANSI A92.2–1969 and this section and to be at least as safe as the equipment was before modification.

(b) (1) Ladder trucks and tower trucks. Aerial ladders shall be secured in the lower traveling position by the locking device on top of the truck cab, and the manually operated device at the base of the ladder before the truck is moved for highway travel.

(2) Extensible and articulating boom platforms.

(i) Lift controls shall be tested each day prior to use to determine that such controls are in safe working condition.
(ii) Only authorized persons shall operate an aerial lift.
(iii) Belting off to an adjacent pole, structure, or equipment while working from an aerial lift shall not be permitted.
(iv) Employees shall always stand firmly on the floor of the basket, and shall not sit or climb on the edge of the basket or use planks, ladders, or other devices for a work position.
(v) A body belt shall be worn and a lanyard attached to the boom or basket when working from an aerial lift. Note to paragraph (b)(2)(v): As of January 1, 1998, subpart M of this part (1926.502(d)) provides that body belts are not acceptable as part of a personal fall arrest system. The use of a body belt in a tethering system or in a restraint system is acceptable and is regulated under 1926.502(e).
(vi) Boom and basket load limits specified by the manufacturer shall not be exceeded.
(vii) The brakes shall be set and when outriggers are used, they shall be positioned on pads or a solid surface. Wheel chocks shall be installed before using an aerial lift on an incline, provided they can be safely installed.
(viii) An aerial lift truck shall not be moved when the boom is elevated in a working position with men in the basket, except for equipment which is specifically designed for this type of operation in accordance with the provisions of paragraphs (a)(1) and (2) of this section.

(ix) Articulating boom and extensible boom platforms, primarily designed as personnel carriers, shall have both platform (upper) and lower controls. Upper controls shall be in or beside the platform within easy reach of the operator. Lower controls shall provide for overriding the upper controls. Controls shall be plainly marked as to their function. Lower level controls shall not be operated unless permission has been obtained from the employee in the lift, except in case of emergency.

(x) Climbers shall not be worn while performing work from an aerial lift.

(xi) The insulated portion of an aerial lift shall not be altered in any manner that might reduce its insulating value.

(xii) Before moving an aerial lift for travel, the boom(s) shall be inspected to see that it is properly cradled and outriggers are in stowed position except as provided in paragraph (b)(2)(viii) of this section.

(3) Electrical tests. All electrical tests shall conform to the requirements of ANSI A92.2–1969 section 5. However equivalent d.c. voltage tests may be used in lieu of the a.c. voltage specified in A92.2–1969; d.c. voltage tests which are approved by the equipment manufacturer or equivalent entity shall be considered an equivalent test for the purpose of this paragraph (b)(3).

(4) Bursting safety factor. The provisions of the American National Standards Institute standard ANSI A92.2–1969, section 4.9 Bursting Safety Factor shall apply to all critical hydraulic and pneumatic components. Critical components are those in which a failure would result in a free fall or free rotation of the boom. All noncritical components shall have a bursting safety factor of at least 2 to 1.

1926 SUBPART M—FALL PROTECTION

Fall, being one of OSHA's Focus Four hazards, is a critical topic for construction project safety management. At the Fire Station 39 project, the highest elevation of the building structure is almost 40 feet above the ground floor. Because many construction activities occurred at elevations above 6 feet, proper fall protection measures were required for the employees engaged in these activities. Chapters 5 to 10 discuss fall-related hazards pertaining to Fire Station 39 and cite selective standards from Subpart M.

1926.500—Scope, Application, and Definitions Applicable To This Subpart

(a) Scope and application.

(1) This subpart sets forth requirements and criteria for fall protection in construction workplaces covered under 29 CFR part 1926. Exception: The provisions of this subpart do not apply when employees are making an inspection, investigation, or assessment of workplace conditions prior to the actual start of construction work or after all construction work has been completed.

(2) Section 1926.501 sets forth those workplaces, conditions, operations, and circumstances for which fall protection shall be provided except as follows:
   (i) Requirements relating to fall protection for employees working on scaffolds are provided in subpart L of this part.
   (ii) Requirements relating to fall protection for employees working on cranes and derricks are provided in subpart CC of this part.
   (iii) Fall protection requirements for employees performing steel erection work (except for towers and tanks) are provided in subpart R of this part.
   (iv) Requirements relating to fall protection for employees working on certain types of equipment used in tunneling operations are provided in subpart S of this part.
   (v) Requirements relating to fall protection for employees engaged in the erection of tanks and communication and broadcast towers are provided in § 1926.105.
   (vi) Requirements relating to fall protection for employees engaged in the construction of electric transmission and distribution lines and equipment are provided in subpart V of this part.
   (vii) Requirements relating to fall protection for employees working on stairways and ladders are provided in subpart X of this part.

(3) Section 1926.502 sets forth the requirements for the installation, construction, and proper use of fall protection required by part 1926, except as follows:
   (i) Performance requirements for guardrail systems used on scaffolds and performance requirements for falling object protection used on scaffolds are provided in subpart L of this part.
   (ii) Performance requirements for stairways, stairrail systems, and handrails are provided in subpart X of this part.
   (iii) Additional performance requirements for personal climbing equipment, lineman's body belts, safety straps, and lanyards are provided in Subpart V of this part.
   (iv) Section 1926.502 does not apply to the erection of tanks and communication and broadcast towers. (Note: Section 1926.104 sets the criteria for body belts, lanyards and lifelines used for fall protection during tank and communication and broadcast tower erection. Paragraphs (b), (c) and (f) of § 1926.107 provide definitions for the pertinent terms.)
   (v) Criteria for steps, handholds, ladders, and grabrails/guardrails/railings required by subpart CC are provided in subpart CC. Sections 1926.502(a), (c) through (e), and (i) apply to activities covered under subpart CC unless otherwise stated in subpart CC. No other paragraphs of § 1926.502 apply to subpart CC.

(4) Section 1926.503 sets forth requirements for training in the installation and use of fall protection systems, except in relation to steel erection activities and the use of equipment covered by subpart CC.

1926.501—Duty To Have Fall Protection

(a) (1) This section sets forth requirements for employers to provide fall protection systems. All fall protection required by this section shall conform to the criteria set forth in 1926.502 of this subpart.

(2) The employer shall determine if the walking/working surfaces on which its employees are to work have the strength and structural integrity to support employees safely. Employees shall be allowed to work on those surfaces only when the surfaces have the requisite strength and structural integrity.

(b) (1) "Unprotected sides and edges." Each employee on a walking/working surface (horizontal and vertical surface) with an unprotected side or edge which is 6 feet (1.8 m) or more above a lower level shall be protected from falling by the use of guardrail systems, safety net systems, or personal fall arrest systems.

(2) "Leading edges."

(i) Each employee who is constructing a leading edge 6 feet (1.8 m) or more above lower levels shall be protected from falling by guardrail systems, safety net systems, or personal fall arrest systems. Exception: When the employer can demonstrate that it is infeasible or creates a greater hazard to use these systems, the employer shall develop and implement a fall protection plan which meets the requirements of paragraph (k) of 1926.502.

Note: There is a presumption that it is feasible and will not create a greater hazard to implement at least one of the above-listed fall protection systems. Accordingly, the employer has the burden of establishing that it is appropriate to implement a fall protection plan which complies with 1926.502(k) for a particular workplace situation, in lieu of implementing any of those systems.

(ii) Each employee on a walking/working surface 6 feet (1.8 m) or more above a lower level where leading edges are under construction, but who is not engaged in the leading edge work, shall be protected from falling by a guardrail system, safety net system, or personal fall arrest system. If a guardrail system is chosen to provide the fall protection, and a controlled access zone has already been established for leading edge work, the control line may be used in lieu of a guardrail along the edge that parallels the leading edge.

(3) "Hoist areas." Each employee in a hoist area shall be protected from falling 6 feet (1.8 m) or more to lower levels by guardrail systems or personal fall arrest systems. If guardrail systems, [or chain, gate, or guardrail] or portions thereof, are removed to facilitate the hoisting operation (e.g., during landing of materials), and an employee must lean through the access opening or out over the edge of the access opening (to receive or guide equipment and materials, for example), that employee shall be protected from fall hazards by a personal fall arrest system.

(4) "Holes."

(i) Each employee on walking/working surfaces shall be protected from falling through holes (including skylights) more than 6 feet (1.8 m) above lower levels, by personal fall arrest systems, covers, or guardrail systems erected around such holes.

(ii) Each employee on a walking/working surface shall be protected from tripping in or stepping into or through holes (including skylights) by covers.

(iii) Each employee on a walking/working surface shall be protected from objects falling through holes (including skylights) by covers.

(5) "Formwork and reinforcing steel." Each employee on the face of formwork or reinforcing steel shall be protected from falling 6 feet (1.8 m) or more to lower levels by personal fall arrest systems, safety net systems, or positioning device systems.

(6) "Ramps, runways, and other walkways." Each employee on ramps, runways, and other walkways shall be protected from falling 6 feet (1.8 m) or more to lower levels by guardrail systems.

(7) "Excavations."

(i) Each employee at the edge of an excavation 6 feet (1.8 m) or more in depth shall be protected from falling by guardrail systems, fences, or barricades when the excavations are not readily seen because of plant growth or other visual barrier;

(ii) Each employee at the edge of a well, pit, shaft, and similar excavation 6 feet (1.8 m) or more in depth shall be protected from falling by guardrail systems, fences, barricades, or covers.

(8) "Dangerous equipment."

(i) Each employee less than 6 feet (1.8 m) above dangerous equipment shall be protected from falling into or onto the dangerous equipment by guardrail systems or by equipment guards.

(ii) Each employee 6 feet (1.8 m) or more above dangerous equipment shall be protected from fall hazards by guardrail systems, personal fall arrest systems, or safety net systems.

(9) "Overhand bricklaying and related work."

(i) Except as otherwise provided in paragraph (b) of this section, each employee performing overhand bricklaying and related work 6 feet (1.8 m) or more above lower levels, shall be protected from falling by guardrail systems, safety net systems, personal fall arrest systems, or shall work in a controlled access zone.

(ii) Each employee reaching more than 10 inches (25 cm) below the level of the walking/working surface on which they are working, shall be protected from falling by a guardrail system, safety net system, or personal fall arrest system.

Note: Bricklaying operations performed on scaffolds are regulated by subpart L—Scaffolds of this part.

(10) "Roofing work on Low-slope roofs." Except as otherwise provided in paragraph (b) of this section, each employee engaged in roofing activities on low-slope roofs, with unprotected sides and edges 6 feet (1.8 m) or more above lower levels shall be protected from falling by guardrail systems, safety net systems, personal fall arrest systems, or a combination of warning line system and guardrail system, warning line system and safety net system, or warning line system and personal fall arrest system, or warning line system and safety monitoring system. Or, on roofs 50-feet (15.25 m) or less in width, the use of a safety monitoring system alone [i.e. without the warning line system] is permitted.

(11) "Steep roofs." Each employee on a steep roof with unprotected sides and edges 6 feet (1.8 m) or more above lower levels shall be protected from falling by guardrail systems with toeboards, safety net systems, or personal fall arrest systems.

(12) "Precast concrete erection." Each employee engaged in the erection of precast concrete members (including, but not limited to the erection of wall panels, columns, beams, and floor and roof "tees") and related operations such as grouting of precast concrete members, who is 6 feet (1.8 m) or more above lower levels shall be protected from falling by guardrail systems, safety net systems, or personal fall arrest systems, unless another provision in paragraph (b) of this section provides for an alternative fall protection measure. Exception: When the employer can demonstrate that it is infeasible or creates a greater hazard to use these systems, the employer shall develop and implement a fall protection plan which meets the requirements of paragraph (k) of 1926.502.

Note: There is a presumption that it is feasible and will not create a greater hazard to implement at least one of the above-listed fall protection systems. Accordingly, the employer has the burden of establishing that it is appropriate to implement a fall protection plan which complies with 1926.502(k) for a particular workplace situation, in lieu of implementing any of those systems.

(13) "Residential construction." Each employee engaged in residential construction activities 6 feet (1.8 m) or more above lower levels shall be protected by guardrail systems, safety net system, or personal fall arrest system unless another provision in paragraph (b) of this section provides for an alternative fall protection measure. Exception: When the employer can demonstrate that it is infeasible or creates a greater hazard to use these systems, the employer shall develop and implement a fall protection plan which meets the requirements of paragraph (k) of 1926.502.

Note: There is a presumption that it is feasible and will not create a greater hazard to implement at least one of the above-listed fall protection systems. Accordingly, the employer has the burden of establishing that it is appropriate to implement a fall protection plan which complies with 1926.502(k) for a particular workplace situation, in lieu of implementing any of those systems.

(14) "Wall openings." Each employee working on, at, above, or near wall openings (including those with chutes attached) where the outside bottom edge of the wall opening is 6 feet (1.8 m) or more above lower levels and the inside bottom edge of the wall opening is less than 39 inches (1.0 m) above the walking/working surface, shall be protected from falling by the use of a guardrail system, a safety net system, or a personal fall arrest system.

(15) "Walking/working surfaces not otherwise addressed." Except as provided in 1926.500(a)(2) or in 1926.501 (b)(1) through (b)(14), each employee on a walking/working surface 6 feet (1.8 m) or more above lower levels shall be protected from falling by a guardrail system, safety net system, or personal fall arrest system.

(c) "Protection from falling objects." When an employee is exposed to falling objects, the employer shall have each employee wear a hard hat and shall implement one of the following measures:

(1) Erect toeboards, screens, or guardrail systems to prevent objects from falling from higher levels; or,

(2) Erect a canopy structure and keep potential fall objects far enough from the edge of the higher level so that those objects would not go over the edge if they were accidentally displaced; or,

(3) Barricade the area to which objects could fall, prohibit employees from entering the barricaded area, and keep objects that may fall far enough away from the edge of a higher level so that those objects would not go over the edge if they were accidentally displaced.

1926.502—Fall Protection Systems Criteria and Practices

Fall hazards due to the use of scaffolds during masonry construction call for the use of guardrail systems. This part of the OSHA standards describes how guardrail systems should be properly set up. Because scaffold systems are extremely common on construction sites, this part of the standard is highly cited.

(a) (1) Fall protection systems required by this part shall comply with the applicable provisions of this section.

(2) Employers shall provide and install all fall protection systems required by this subpart for an employee, and shall comply with all other pertinent requirements of this subpart before that employee begins the work that necessitates the fall protection.

(b) Guardrail systems and their use shall comply with the following provisions:

(1) Top edge height of top rails, or equivalent guardrail system members, shall be 42 inches (1.1 m) plus or minus 3 inches (8 cm) above the walking/working level. When conditions warrant, the height of the top edge may exceed the 45-inch height, provided the guardrail system meets all other criteria of this paragraph.

Note: When employees are using stilts, the top edge height of the top rail, or equivalent member, shall be increased an amount equal to the height of the stilts.

(2) Midrails, screens, mesh, intermediate vertical members, or equivalent intermediate structural members shall be installed between the top edge of the guardrail system and the walking/working surface when there is no wall or parapet wall at least 21 inches (53 cm) high.

(i) Midrails, when used, shall be installed at a height midway between the top edge of the guardrail system and the walking/working level.

(ii) Screens and mesh, when used, shall extend from the top rail to the walking/working level and along the entire opening between top rail supports.

(iii) Intermediate members (such as balusters), when used between posts, shall be not more than 19 inches (48 cm) apart.

(iv) Other structural members (such as additional midrails and architectural panels) shall be installed such that there are no openings in the guardrail system that are more than 19 inches (.5 m) wide.

(3) Guardrail systems shall be capable of withstanding, without failure, a force of at least 200 pounds (890 N) applied within 2 inches (5.1 cm) of the top edge, in any outward or downward direction, at any point along the top edge.

(4) When the 200 pound (890 N) test load specified in paragraph (b)(3) of this section is applied in a downward direction, the top edge of the guardrail shall not deflect to a height less than 39 inches (1.0 m) above the walking/working level.

(5) Midrails, screens, mesh, intermediate vertical members, solid panels, and equivalent structural members shall be capable of withstanding, without failure, a force of at least 150 pounds (666 N) applied in any downward or outward direction at any point along the midrail or other member.

(6) Guardrail systems shall be so surfaced as to prevent injury to an employee from punctures or lacerations, and to prevent snagging of clothing.

(7) The ends of all top rails and midrails shall not overhang the terminal posts, except where such overhang does not constitute a projection hazard.

(8) Steel banding and plastic banding shall not be used as top rails or midrails.

(9) Top rails and midrails shall be at least one-quarter inch (0.6 cm) nominal diameter or thickness to prevent cuts and lacerations. If wire rope is used for top rails, it shall be flagged at not more than 6-foot intervals with high-visibility material.

(10) When guardrail systems are used at hoisting areas, a chain, gate or removable guardrail section shall be placed across the access opening between guardrail sections when hoisting operations are not taking place.

(11) When guardrail systems are used at holes, they shall be erected on all unprotected sides or edges of the hole.

(12) When guardrail systems are used around holes used for the passage of materials, the hole shall have not more than two sides provided with removable guardrail sections to allow the passage of materials. When the hole is not in use, it shall be closed over with a cover, or a guardrail system shall be provided along all unprotected sides or edges.

(13) When guardrail systems are used around holes which are used as points of access (such as ladderways), they shall be provided with a gate, or be so offset that a person cannot walk directly into the hole.

(14) Guardrail systems used on ramps and runways shall be erected along each unprotected side or edge.

(15) Manila, plastic or synthetic rope being used for top rails or midrails shall be inspected as frequently as necessary to ensure that it continues to meet the strength requirements of paragraph (b)(3) of this section.

(c) Safety net systems and their use shall comply with the following provisions:

(1) Safety nets shall be installed as close as practicable under the walking/working surface on which employees are working, but in no case more than 30 feet (9.1 m) below such level. When nets are used on bridges, the potential fall area from the walking/working surface to the net shall be unobstructed.

(2) Safety nets shall extend outward from the outermost projection of the work surface as follows:

Vertical Distance from Working Level To Horizontal Plane of Net	Minimum Required Horizontal Distance of Outer Edge of Net From the Edge of the Working Surface
Up to 5 feet	8 feet
More than 5 feet up to 10 feet	10 feet
More than 10 feet	13 feet

(3) Safety nets shall be installed with sufficient clearance under them to prevent contact with the surface or structures below when subjected to an impact force equal to the drop test specified in paragraph (c)(4) of this section.

(4) Safety nets and their installations shall be capable of absorbing an impact force equal to that produced by the drop test specified in paragraph (c)(4)(i) of this section.

(i) Except as provided in paragraph (c)(4)(ii) of this section, safety nets and safety net installations shall be drop-tested at the jobsite after initial installation and before being used as a fall protection system, whenever relocated, after major repair, and at 6-month intervals if left in one place. The drop-test shall consist of a 400 pound (180 kg) bag of sand 30 + or − 2 inches (76 + or − 5 cm) in diameter dropped into the net from the highest walking/working surface at which employees are exposed to fall hazards, but not from less than 42 inches (1.1 m) above that level.

(ii) When the employer can demonstrate that it is unreasonable to perform the drop-test required by paragraph (c)(4)(i) of this section, the employer (or a designated competent person) shall certify that the net and net installation is in compliance with the provisions of paragraphs (c)(3) and (c)(4)(i) of this section by preparing a certification record prior to the net being used as a fall protection system. The certification record must include an identification of the net and net installation for which the certification record is being prepared; the date that it was determined that the identified net and net installation were in compliance with paragraph (c)(3) of this section and the signature of the person making the determination and certification. The most recent certification record for each net and net installation shall be available at the jobsite for inspection.

(5) Defective nets shall not be used. Safety nets shall be inspected at least once a week for wear, damage, and other deterioration. Defective components shall be removed from service. Safety nets shall also be inspected after any occurrence which could affect the integrity of the safety net system.

(6) Materials, scrap pieces, equipment, and tools which have fallen into the safety net shall be removed as soon as possible from the net and at least before the next work shift.

(7) The maximum size of each safety net mesh opening shall not exceed 36 square inches (230 cm) nor be longer than 6 inches (15 cm) on any side, and the opening, measured center-to-center of mesh ropes or webbing, shall not be longer than 6 inches (15 cm). All mesh crossings shall be secured to prevent enlargement of the mesh opening.

(8) Each safety net (or section of it) shall have a border rope for webbing with a minimum breaking strength of 5,000 pounds (22.2 kN).

(9) Connections between safety net panels shall be as strong as integral net components and shall be spaced not more than 6 inches (15 cm) apart.

(d) Personal fall arrest systems and their use shall comply with the provisions set forth below. Effective January 1, 1998, body belts are not acceptable as part of a personal fall arrest system. Note: The use of a body belt in a positioning device system is acceptable and is regulated under paragraph (e) of this section.

(1) Connectors shall be drop forged, pressed or formed steel, or made of equivalent materials.

(2) Connectors shall have a corrosion-resistant finish, and all surfaces and edges shall be smooth to prevent damage to interfacing parts of the system.

(3) Dee-rings and snaphooks shall have a minimum tensile strength of 5,000 pounds (22.2 kN).

(4) Dee-rings and snaphooks shall be proof-tested to a minimum tensile load of 3,600 pounds (16 kN) without cracking, breaking, or taking permanent deformation.

(5) Snaphooks shall be sized to be compatible with the member to which they are connected to prevent unintentional disengagement of the snaphook by depression of the snaphook keeper by the connected member, or shall be a locking type snaphook designed and used to prevent disengagement of the snaphook by the contact of the snaphook keeper by the connected member. Effective January 1, 1998, only locking type snaphooks shall be used.

(6) Unless the snaphook is a locking type and designed for the following connections, snaphooks shall not be engaged:
(i) directly to webbing, rope or wire rope;
(ii) to each other;
(iii) to a dee-ring to which another snaphook or other connector is attached;
(iv) to a horizontal lifeline; or
(v) to any object which is incompatibly shaped or dimensioned in relation to the snaphook such that unintentional disengagement could occur by the connected object being able to depress the snaphook keeper and release itself.

(7) On suspended scaffolds or similar work platforms with horizontal lifelines which may become vertical lifelines, the devices used to connect to a horizontal lifeline shall be capable of locking in both directions on the lifeline.

(8) Horizontal lifelines shall be designed, installed, and used, under the supervision of a qualified person, as part of a complete personal fall arrest system, which maintains a safety factor of at least two.

(9) Lanyards and vertical lifelines shall have a minimum breaking strength of 5,000 pounds (22.2 kN).

(10) (i) Except as provided in paragraph (d)(10)(ii) of this section, when vertical lifelines are used, each employee shall be attached to a separate lifeline.
(ii) During the construction of elevator shafts, two employees may be attached to the same lifeline in the hoistway, provided both employees are working atop a false car that is equipped with guardrails; the strength of the lifeline is 10,000 pounds [5,000 pounds per employee attached] (44.4 kN); and all other criteria specified in this paragraph for lifelines have been met.

(11) Lifelines shall be protected against being cut or abraded.

(12) Self-retracting lifelines and lanyards which automatically limit free fall distance to 2 feet (0.61 m) or less shall be capable of sustaining a minimum tensile load of 3,000 pounds (13.3 kN) applied to the device with the lifeline or lanyard in the fully extended position.

(13) Self-retracting lifelines and lanyards which do not limit free fall distance to 2 feet (0.61 m) or less, ripstitch lanyards, and tearing and deforming lanyards shall be capable of sustaining a minimum tensile load of 5,000 pounds (22.2 kN) applied to the device with the lifeline or lanyard in the fully extended position.

(14) Ropes and straps (webbing) used in lanyards, lifelines, and strength components of body belts and body harnesses shall be made from synthetic fibers.

(15) Anchorages used for attachment of personal fall arrest equipment shall be independent of any anchorage being used to support or suspend platforms and capable of supporting at least 5,000 pounds

(22.2 kN) per employee attached, or shall be designed, installed, and used as follows:

(i) as part of a complete personal fall arrest system which maintains a safety factor of at least two; and

(ii) under the supervision of a qualified person.

(16) Personal fall arrest systems, when stopping a fall, shall:

(i) limit maximum arresting force on an employee to 900 pounds (4 kN) when used with a body belt;

(ii) limit maximum arresting force on an employee to 1,800 pounds (8 kN) when used with a body harness;

(iii) be rigged such that an employee can neither free fall more than 6 feet (1.8 m), nor contact any lower level;

(iv) bring an employee to a complete stop and limit maximum deceleration distance an employee travels to 3.5 feet (1.07 m); and,

(v) have sufficient strength to withstand twice the potential impact energy of an employee free falling a distance of 6 feet (1.8 m), or the free fall distance permitted by the system, whichever is less.

Note: If the personal fall arrest system meets the criteria, and if the system is being used by an employee having a combined person and tool weight of less than 310 pounds (140 kg), the system will be considered to be in compliance with the provisions of paragraph (d)(16) of this section. If the system is used by an employee having a combined tool and body weight of 310 pounds (140 kg) or more, then the employer must appropriately modify the criteria and protocols of the Appendix to provide proper protection for such heavier weights, or the system will not be deemed to be in compliance with the requirements of paragraph (d)(16) of this section.

(17) The attachment point of the body belt shall be located in the center of the wearer's back. The attachment point of the body harness shall be located in the center of the wearer's back near shoulder level, or above the wearer's head.

(18) Body belts, harnesses, and components shall be used only for employee protection (as part of a personal fall arrest system or positioning device system) and not to hoist materials.

(19) Personal fall arrest systems and components subjected to impact loading shall be immediately removed from service and shall not be used again for employee protection until inspected and determined by a competent person to be undamaged and suitable for reuse.

(20) The employer shall provide for prompt rescue of employees in the event of a fall or shall assure that employees are able to rescue themselves.

(21) Personal fall arrest systems shall be inspected prior to each use for wear, damage and other deterioration, and defective components shall be removed from service.

(22) Body belts shall be at least one and five-eighths ($1\frac{5}{8}$) inches (4.1 cm) wide.

(23) Personal fall arrest systems shall not be attached to guardrail systems, nor shall they be attached to hoists except as specified in other subparts of this Part.

(24) When a personal fall arrest system is used at hoist areas, it shall be rigged to allow the movement of the employee only as far as the edge of the walking/working surface.

(e) "Positioning device systems." Positioning device systems and their use shall conform to the following provisions:

(1) Positioning devices shall be rigged such that an employee cannot free fall more than 2 feet (.6 m).

(2) Positioning devices shall be secured to an anchorage capable of supporting at least twice the potential impact load of an employee's fall or 3,000 pounds (13.3 kN), whichever is greater.

(3) Connectors shall be drop forged, pressed or formed steel, or made of equivalent materials.

(4) Connectors shall have a corrosion-resistant finish, and all surfaces and edges shall be smooth to prevent damage to interfacing parts of this system.

(5) Connecting assemblies shall have a minimum tensile strength of 5,000 pounds (22.2 kN).

(6) Dee-rings and snaphooks shall be proof-tested to a minimum tensile load of 3,600 pounds (16 kN) without cracking, breaking, or taking permanent deformation.

(7) Snaphooks shall be sized to be compatible with the member to which they are connected to prevent unintentional disengagement of the snaphook by depression of the snaphook keeper by the connected member, or shall be a locking type snaphook designed and used to prevent disengagement of the snaphook by the contact of the snaphook keeper by the connected member. As of January 1, 1998, only locking type snaphooks shall be used.

(8) Unless the snaphook is a locking type and designed for the following connections, snaphooks shall not be engaged:

(i) directly to webbing, rope or wire rope;

(ii) to each other;

(iii) to a dee-ring to which another snaphook or other connector is attached;

(iv) to a horizontal lifeline; or

(v) to any object which is incompatibly shaped or dimensioned in relation to the snaphook such that unintentional disengagement could occur by the connected object being able to depress the snaphook keeper and release itself.

(9) Positioning device systems shall be inspected prior to each use for wear, damage, and other deterioration, and defective components shall be removed from service.

(10) Body belts, harnesses, and components shall be used only for employee protection (as part of a personal fall arrest system or positioning device system) and not to hoist materials.

(f) "Warning line systems." Warning line systems [See 1926.501(b)(10)] and their use shall comply with the following provisions:

(1) The warning line shall be erected around all sides of the roof work area.

(i) When mechanical equipment is not being used, the warning line shall be erected not less than 6 feet (1.8 m) from the roof edge.

(ii) When mechanical equipment is being used, the warning line shall be erected not less than 6 feet (1.8 m) from the roof edge which is parallel to the direction of mechanical equipment operation, and not less than 10 feet (3.1 m) from the roof edge which is perpendicular to the direction of mechanical equipment operation.

(iii) Points of access, materials handling areas, storage areas, and hoisting areas shall be connected to the work area by an access path formed by two warning lines.

(iv) When the path to a point of access is not in use, a rope, wire, chain, or other barricade, equivalent in strength and height to the warning line, shall be placed across the path at the point where the path intersects the warning line erected around the work area, or the path shall be offset such that a person cannot walk directly into the work area.

(2) Warning lines shall consist of ropes, wires, or chains, and supporting stanchions erected as follows:

(i) The rope, wire, or chain shall be flagged at not more than 6-foot (1.8 m) intervals with high-visibility material;

(ii) The rope, wire, or chain shall be rigged and supported in such a way that its lowest point (including sag) is no less than 34 inches (.9 m) from the walking/working surface and its highest point is no more than 39 inches (1.0 m) from the walking/working surface;

(iii) After being erected, with the rope, wire, or chain attached, stanchions shall be capable of resisting, without tipping over, a force of at least 16 pounds (71 N) applied horizontally against the stanchion, 30 inches (.8 m) above the walking/working surface, perpendicular to the warning line, and in the direction of the floor, roof, or platform edge;

(iv) The rope, wire, or chain shall have a minimum tensile strength of 500 pounds (2.22 kN), and after being attached to the stanchions, shall be capable of supporting, without breaking, the loads applied to the stanchions as prescribed in paragraph (f)(2)(iii) of this section; and

(v) The line shall be attached at each stanchion in such a way that pulling on one section of the line between stanchions will not result in slack being taken up in adjacent sections before the stanchion tips over.

(3) No employee shall be allowed in the area between a roof edge and a warning line unless the employee is performing roofing work in that area.

(4) Mechanical equipment on roofs shall be used or stored only in areas where employees are protected by a warning line system, guardrail system, or personal fall arrest system.

(g) Controlled access zones [See 1926.501(b)(9) and 1926.502(k)] and their use shall conform to the following provisions.

(1) When used to control access to areas where leading edge and other operations are taking place the controlled access zone shall be defined by a control line or by any other means that restricts access.

(i) When control lines are used, they shall be erected not less than 6 feet (1.8 m) nor more than 25 feet (7.7 m) from the unprotected or leading edge, except when erecting precast concrete members.

(ii) When erecting precast concrete members, the control line shall be erected not less than 6 feet (1.8 m) nor more than 60 feet (18 m) or half the length of the member being erected, whichever is less, from the leading edge.

(iii) The control line shall extend along the entire length of the unprotected or leading edge and shall be approximately parallel to the unprotected or leading edge.

(iv) The control line shall be connected on each side to a guardrail system or wall.

(2) When used to control access to areas where overhand bricklaying and related work are taking place:

(i) The controlled access zone shall be defined by a control line erected not less than 10 feet (3.1 m) nor more than 15 feet (4.5 m) from the working edge.

(ii) The control line shall extend for a distance sufficient for the controlled access zone to enclose all employees performing overhand bricklaying and related work at the working edge and shall be approximately parallel to the working edge.

(iii) Additional control lines shall be erected at each end to enclose the controlled access zone.

(iv) Only employees engaged in overhand bricklaying or related work shall be permitted in the controlled access zone.

(3) Control lines shall consist of ropes, wires, tapes, or equivalent materials, and supporting stanchions as follows:

(i) Each line shall be flagged or otherwise clearly marked at not more than 6-foot (1.8 m) intervals with high-visibility material.

(ii) Each line shall be rigged and supported in such a way that its lowest point (including sag) is not less than 39 inches (1 m) from the walking/working surface and its highest point is not more than 45 inches (1.3 m) [50 inches (1.3 m) when overhand bricklaying operations are being performed] from the walking/working surface.

(iii) Each line shall have a minimum breaking strength of 200 pounds (.88 kN).

(4) On floors and roofs where guardrail systems are not in place prior to the beginning of overhand bricklaying operations, controlled access zones shall be enlarged, as necessary, to enclose all points of access, material handling areas, and storage areas.

(5) On floors and roofs where guardrail systems are in place, but need to be removed to allow overhand bricklaying work or leading edge work to take place, only that portion of the guardrail necessary to accomplish that day's work shall be removed.

(h) Safety monitoring systems [See 1926.501(b)(10) and 1926.502(k)] and their use shall comply with the following provisions:

(1) The employer shall designate a competent person to monitor the safety of other employees and the employer shall ensure that the safety monitor complies with the following requirements:

(i) The safety monitor shall be competent to recognize fall hazards;

(ii) The safety monitor shall warn the employee when it appears that the employee is unaware of a fall hazard or is acting in an unsafe manner;

(iii) The safety monitor shall be on the same walking/working surface and within visual sighting distance of the employee being monitored;

(iv) The safety monitor shall be close enough to communicate orally with the employee; and

(v) The safety monitor shall not have other responsibilities which could take the monitor's attention from the monitoring function.

(2) Mechanical equipment shall not be used or stored in areas where safety monitoring systems are being used to monitor employees engaged in roofing operations on low-slope roofs.

(3) No employee, other than an employee engaged in roofing work [on low-sloped roofs] or an employee covered by a fall protection plan, shall be allowed in an area where an employee is being protected by a safety monitoring system.

(4) Each employee working in a controlled access zone shall be directed to comply promptly with fall hazard warnings from safety monitors.

(i) Covers for holes in floors, roofs, and other walking/working surfaces shall meet the following requirements:

(1) Covers located in roadways and vehicular aisles shall be capable of supporting, without failure, at least twice the maximum axle load of the largest vehicle expected to cross over the cover.

(2) All other covers shall be capable of supporting, without failure, at least twice the weight of employees, equipment, and materials that may be imposed on the cover at any one time.

(3) All covers shall be secured when installed so as to prevent accidental displacement by the wind, equipment, or employees.

(4) All covers shall be color coded or they shall be marked with the word "HOLE" or "COVER" to provide warning of the hazard.

Note: This provision does not apply to cast iron manhole covers or steel grates used on streets or roadways.

(j) Falling object protection shall comply with the following provisions:

(1) Toeboards, when used as falling object protection, shall be erected along the edge of the overhead walking/working surface for a distance sufficient to protect employees below.

(2) Toeboards shall be capable of withstanding, without failure, a force of at least 50 pounds (222 N) applied in any downward or outward direction at any point along the toeboard.

(3) Toeboards shall be a minimum of 3½ inches (9 cm) in vertical height from their top edge to the level of the walking/working surface. They shall have not more than ¼ inch (0.6 cm) clearance above the walking/working surface. They shall be solid or have openings not over 1 inch (2.5 cm) in greatest dimension.

(4) Where tools, equipment, or materials are piled higher than the top edge of a toeboard, paneling or screening shall be erected from the walking/working surface or toeboard to the top of a guardrail system's top rail or midrail, for a distance sufficient to protect employees below.

(5) Guardrail systems, when used as falling object protection, shall have all openings small enough to prevent passage of potential falling objects.

(6) During the performance of overhand bricklaying and related work:

(i) No materials or equipment except masonry and mortar shall be stored within 4 feet (1.2 m) of the working edge.

(ii) Excess mortar, broken or scattered masonry units, and all other materials and debris shall be kept clear from the work area by removal at regular intervals.

(7) During the performance of roofing work:

(i) Materials and equipment shall not be stored within 6 feet (1.8 m) of a roof edge unless guardrails are erected at the edge.

(ii) Materials which are piled, grouped, or stacked near a roof edge shall be stable and self-supporting.

(8) Canopies, when used as falling object protection, shall be strong enough to prevent collapse and to prevent penetration by any objects which may fall onto the canopy.

(k) "Fall protection plan." This option is available only to employees engaged in leading edge work, precast concrete erection work, or residential construction work (See 1926.501(b)(2), (b)(12), and (b)(13)) who can demonstrate that it is infeasible or it creates a greater hazard to use conventional fall protection equipment. The fall protection plan must conform to the following provisions.

(1) The fall protection plan shall be prepared by a qualified person and developed specifically for the site where the leading edge work, precast concrete work, or residential construction work is being performed and the plan must be maintained up to date.

(2) Any changes to the fall protection plan shall be approved by a qualified person.

(3) A copy of the fall protection plan with all approved changes shall be maintained at the job site.

(4) The implementation of the fall protection plan shall be under the supervision of a competent person.

(5) The fall protection plan shall document the reasons why the use of conventional fall protection systems (guardrail systems, personal fall arrest systems, or safety nets systems) are infeasible or why their use would create a greater hazard.

(6) The fall protection plan shall include a written discussion of other measures that will be taken to reduce or eliminate the fall hazard for workers who cannot be provided with protection from the conventional fall protection systems. For example, the employer shall discuss the extent to which scaffolds, ladders, or vehicle mounted work platforms can be used to provide a safer working surface and thereby reduce the hazard of falling.

(7) The fall protection plan shall identify each location where conventional fall protection methods cannot be used. These locations shall then be classified as controlled access zones and the employer must comply with the criteria in paragraph (g) of this section.

(8) Where no other alternative measure has been implemented, the employer shall implement a safety monitoring system in conformance with 1926.502(h).

(9) The fall protection plan must include a statement which provides the name or other method of identification for each employee who is designated to work in controlled access zones. No other employees may enter controlled access zones.

(10) In the event an employee falls, or some other related, serious incident occurs, (e.g., a near miss) the employer shall investigate the circumstances of the fall or other incident to determine if the fall protection plan needs to be changed (e.g. new practices, procedures, or training) and shall implement those changes to prevent similar types of falls or incidents.

1926 SUBPART P—EXCAVATIONS

Chapter 5 discusses Fire Station 39's foundation excavation and how 1926 Subpart P requires the installation of protection systems to avoid cave-ins during trench excavation.

1926.650—Scope, Application, and Definitions Applicable To This Subpart

(a) Scope and application. This subpart applies to all open excavations made in the earth's surface. Excavations are defined to include trenches.

1926.651—Specific Excavation Requirements

(a) All surface encumbrances that are located so as to create a hazard to employees shall be removed or supported, as necessary, to safeguard employees.

(b) Underground installations.

(1) The estimated location of utility installations, such as sewer, telephone, fuel, electric, water lines, or any other underground installations that

reasonably may be expected to be encountered during excavation work, shall be determined prior to opening an excavation.

(2) Utility companies or owners shall be contacted within established or customary local response times, advised of the proposed work, and asked to establish the location of the utility underground installations prior to the start of actual excavation. When utility companies or owners cannot respond to a request to locate underground utility installations within 24 hours (unless a longer period is required by state or local law), or cannot establish the exact location of these installations, the employer may proceed, provided the employer does so with caution, and provided detection equipment or other acceptable means to locate utility installations are used.

(3) When excavation operations approach the estimated location of underground installations, the exact location of the installations shall be determined by safe and acceptable means.

(4) While the excavation is open, underground installations shall be protected, supported or removed as necessary to safeguard employees.

(c) Access and egress.

(1) Structural ramps.

(i) Structural ramps that are used solely by employees as a means of access or egress from excavations shall be designed by a competent person. Structural ramps used for access or egress of equipment shall be designed by a competent person qualified in structural design, and shall be constructed in accordance with the design.

(ii) Ramps and runways constructed of two or more structural members shall have the structural members connected together to prevent displacement.

(iii) Structural members used for ramps and runways shall be of uniform thickness.

(iv) Cleats or other appropriate means used to connect runway structural members shall be attached to the bottom of the runway or shall be attached in a manner to prevent tripping.

(v) Structural ramps used in lieu of steps shall be provided with cleats or other surface treatments on the top surface to prevent slipping.

(2) Means of egress from trench excavations. A stairway, ladder, ramp or other safe means of egress shall be located in trench excavations that are 4 feet (1.22 m) or more in depth so as to require no more than 25 feet (7.62 m) of lateral travel for employees.

(d) Exposure to vehicular traffic. Employees exposed to public vehicular traffic shall be provided with, and shall wear, warning vests or other suitable garments marked with or made of reflectorized or high-visibility material.

(e) Exposure to falling loads. No employee shall be permitted underneath loads handled by lifting or digging equipment. Employees shall be required to stand away from any vehicle being loaded or unloaded to avoid being struck by any spillage or falling materials. Operators may remain in the cabs of vehicles being loaded or unloaded when the vehicles are equipped, in accordance with 1926.601(b)(6), to provide adequate protection for the operator during loading and unloading operations.

(f) Warning system for mobile equipment. When mobile equipment is operated adjacent to an excavation, or when such equipment is required to approach the edge of an excavation, and the operator does not have a clear and direct view of the edge of the excavation, a warning system shall be utilized such as barricades, hand or mechanical signals, or stop logs. If possible, the grade should be away from the excavation.

(g) Hazardous atmospheres.

(1) Testing and controls. In addition to the requirements set forth in subparts D and E of this part (29 CFR 1926.50—1926.107) to prevent exposure to harmful levels of atmospheric contaminants and to assure acceptable atmospheric conditions, the following requirements shall apply:

(i) Where oxygen deficiency (atmospheres containing less than 19.5 percent oxygen) or a hazardous atmosphere exists or could reasonably be expected to exist, such as in excavations in landfill areas or excavations in areas where hazardous substances are stored nearby, the atmospheres in the excavation shall be tested before employees enter excavations greater than 4 feet (1.22 m) in depth.

(ii) Adequate precautions shall be taken to prevent employee exposure to atmospheres containing less than 19.5 percent oxygen and other hazardous atmospheres. These precautions include providing proper respiratory protection or ventilation in accordance with subparts D and E of this part respectively.

(iii) Adequate precaution shall be taken such as providing ventilation, to prevent employee exposure to an atmosphere containing a concentration of a flammable gas in excess of 20 percent of the lower flammable limit of the gas.

(iv) When controls are used that are intended to reduce the level of atmospheric contaminants to acceptable levels, testing shall be conducted as often as necessary to ensure that the atmosphere remains safe.

(2) Emergency rescue equipment.

(i) Emergency rescue equipment, such as breathing apparatus, a safety harness and line, or a basket stretcher, shall be readily available where hazardous atmospheric conditions exist or may reasonably be expected to develop during work in an excavation. This equipment shall be attended when in use.

(ii) Employees entering bell-bottom pier holes, or other similar deep and confined footing excavations, shall wear a harness with a lifeline securely attached to it. The lifeline shall be separate from any line used to handle materials, and shall be individually attended at all times while the employee wearing the lifeline is in the excavation.

(h) Protection from hazards associated with water accumulation.

(1) Employees shall not work in excavations in which there is accumulated water, or in excavations in which water is accumulating, unless adequate

precautions have been taken to protect employees against the hazards posed by water accumulation. The precautions necessary to protect employees adequately vary with each situation, but could include special support or shield systems to protect from cave-ins, water removal to control the level of accumulating water, or use of a safety harness and lifeline.

(2) If water is controlled or prevented from accumulating by the use of water removal equipment, the water removal equipment and operations shall be monitored by a competent person to ensure proper operation.

(3) If excavation work interrupts the natural drainage of surface water (such as streams), diversion ditches, dikes, or other suitable means shall be used to prevent surface water from entering the excavation and to provide adequate drainage of the area adjacent to the excavation. Excavations subject to runoff from heavy rains will require an inspection by a competent person and compliance with paragraphs (h)(1) and (h)(2) of this section.

(i) Stability of adjacent structures.

(1) Where the stability of adjoining buildings, walls, or other structures is endangered by excavation operations, support systems such as shoring, bracing, or underpinning shall be provided to ensure the stability of such structures for the protection of employees.

(2) Excavation below the level of the base or footing of any foundation or retaining wall that could be reasonably expected to pose a hazard to employees shall not be permitted except when:

(i) A support system, such as underpinning, is provided to ensure the safety of employees and the stability of the structure; or

(ii) The excavation is in stable rock; or

(iii) A registered professional engineer has approved the determination that the structure is sufficiently removed from the excavation so as to be unaffected by the excavation activity; or

(iv) A registered professional engineer has approved the determination that such excavation work will not pose a hazard to employees.

(3) Sidewalks, pavements and appurtenant structure shall not be undermined unless a support system or another method of protection is provided to protect employees from the possible collapse of such structures.

(j) Protection of employees from loose rock or soil.

(1) Adequate protection shall be provided to protect employees from loose rock or soil that could pose a hazard by falling or rolling from an excavation face. Such protection shall consist of scaling to remove loose material; installation of protective barricades at intervals as necessary on the face to stop and contain falling material; or other means that provide equivalent protection.

(2) Employees shall be protected from excavated or other materials or equipment that could pose a hazard by falling or rolling into excavations. Protection shall be provided by placing and keeping such materials or equipment at least 2 feet (.61 m) from the edge of excavations, or by the use of retaining devices that are sufficient to prevent materials or equipment from falling or rolling into excavations, or by a combination of both if necessary.

(k) Inspections.

(1) Daily inspections of excavations, the adjacent areas, and protective systems shall be made by a competent person for evidence of a situation that could result in possible cave-ins, indications of failure of protective systems, hazardous atmospheres, or other hazardous conditions. An inspection shall be conducted by the competent person prior to the start of work and as needed throughout the shift. Inspections shall also be made after every rainstorm or other hazard increasing occurrence. These inspections are only required when employee exposure can be reasonably anticipated.

(2) Where the competent person finds evidence of a situation that could result in a possible cave-in, indications of failure of protective systems, hazardous atmospheres, or other hazardous conditions, exposed employees shall be removed from the hazardous area until the necessary precautions have been taken to ensure their safety.

(l) Walkways shall be provided where employees or equipment are required or permitted to cross over excavations. Guardrails which comply with 1926.502(b) shall be provided where walkways are 6 feet (1.8 m) or more above lower levels.

1926.652—Requirements for Protective Systems

(a) Protection of employees in excavations.

(1) Each employee in an excavation shall be protected from cave-ins by an adequate protective system designed in accordance with paragraph (b) or (c) of this section except when:

(i) Excavations are made entirely in stable rock; or

(ii) Excavations are less than 5 feet (1.52 m) in depth and examination of the ground by a competent person provides no indication of a potential cave-in.

(2) Protective systems shall have the capacity to resist without failure all loads that are intended or could reasonably be expected to be applied or transmitted to the system.

(b) Design of sloping and benching systems. The slopes and configurations of sloping and benching systems shall be selected and constructed by the employer or his designee and shall be in accordance with the requirements of paragraph (b)(1); or, in the alternative, paragraph (b)(2); or, in the alternative, paragraph (b)(3); or, in the alternative, paragraph (b)(4), as follows:

(1) Option (1)—Allowable configurations and slopes. Excavations shall be sloped at an angle not steeper than one and one-half horizontal to one vertical (34 degrees measured from the horizontal), unless the employer uses one of the other options listed below.

(2) Option (2)—Designs using other tabulated data.

(i) Designs of sloping or benching systems shall be selected from and in accordance with tabulated data, such as tables and charts.

(ii) The tabulated data shall be in written form and shall include all of the following:

(A) Identification of the parameters that affect the selection of a sloping or benching system drawn from such data;

(B) Identification of the limits of use of the data, to include the magnitude and configuration of slopes determined to be safe;

(C) Explanatory information as may be necessary to aid the user in making a correct selection of a protective system from the data.

(iii) At least one copy of the tabulated data which identifies the registered professional engineer who approved the data, shall be maintained at the jobsite during construction of the protective system. After that time the data may be stored off the jobsite, but a copy of the data shall be made available to the Secretary upon request.

(3) Option (3)—Design by a registered professional engineer.

(i) Sloping and benching systems not utilizing Option (1) or Option (2) under paragraph (b) of this section shall be approved by a registered professional engineer.

(ii) Designs shall be in written form and shall include at least the following:

(A) The magnitude of the slopes that were determined to be safe for the particular project;

(B) The configurations that were determined to be safe for the particular project;

(C) The identity of the registered professional engineer approving the design.

(iii) At least one copy of the design shall be maintained at the jobsite while the slope is being constructed. After that time the design need not be at the jobsite, but a copy shall be made available to the Secretary upon request.

(c) Design of support systems, shield systems, and other protective systems. Designs of support systems, shield systems, and other protective systems shall be selected and constructed by the employer or his designee and shall be in accordance with the requirements of paragraph (c)(1); or, in the alternative, paragraph (c)(2); or, in the alternative, paragraph (c)(3); as follows:

(1) Option (1)—Designs using manufacturer's tabulated data.

(i) Design of support systems, shield systems, or other protective systems that are drawn from manufacturer's tabulated data shall be in accordance with all specifications, recommendations, and limitations issued or made by the manufacturer.

(ii) Deviation from the specifications, recommendations, and limitations issued or made by the manufacturer shall only be allowed after the manufacturer issues specific written approval.

(iii) Manufacturer's specifications, recommendations, and limitations, and manufacturer's approval to deviate from the specifications, recommendations, and limitations shall be in written form at the jobsite during construction of the protective system. After that time this data may be stored off the jobsite, but a copy shall be made available to the Secretary upon request.

(2) Option (2)—Designs using other tabulated data.

(i) Designs of support systems, shield systems, or other protective systems shall be selected from and be in accordance with tabulated data, such as tables and charts.

(ii) The tabulated data shall be in written form and include all of the following:

(A) Identification of the parameters that affect the selection of a protective system drawn from such data;

(B) Identification of the limits of use of the data;

(C) Explanatory information as may be necessary to aid the user in making a correct selection of a protective system from the data.

(iii) At least one copy of the tabulated data, which identifies the registered professional engineer who approved the data, shall be maintained at the jobsite during construction of the protective system. After that time the data may be stored off the jobsite, but a copy of the data shall be made available to the Secretary upon request.

(3) Option (3)—Design by a registered professional engineer.

(i) Support systems, shield systems, and other protective systems not utilizing Option 1, Option 2 or Option 3, above, shall be approved by a registered professional engineer.

(ii) Designs shall be in written form and shall include the following:

(A) A plan indicating the sizes, types, and configurations of the materials to be used in the protective system; and

(B) The identity of the registered professional engineer approving the design.

(iii) At least one copy of the design shall be maintained at the jobsite during construction of the protective system. After that time, the design may be stored off the jobsite, but a copy of the design shall be made available to the Secretary upon request.

(d) Materials and equipment.

(1) Materials and equipment used for protective systems shall be free from damage or defects that might impair their proper function.

(2) Manufactured materials and equipment used for protective systems shall be used and maintained in a manner that is consistent with the recommendations of the manufacturer, and in a manner that will prevent employee exposure to hazards.

(3) When material or equipment that is used for protective systems is damaged, a competent person shall examine the material or equipment and evaluate its suitability for continued use. If the competent person cannot assure the material or equipment is able to support the intended loads or is otherwise suitable for safe use, then such material or equipment shall be removed from service, and shall be evaluated and approved by a registered professional engineer before being returned to service.

(e) Installation and removal of support.

(1) General.

(i) Members of support systems shall be securely connected together to prevent sliding, falling, kickouts, or other predictable failure.

(ii) Support systems shall be installed and removed in a manner that protects employees from cave-ins, structural collapses, or from being struck by members of the support system.

(iii) Individual members of support systems shall not be subjected to loads exceeding those which those members were designed to withstand.

(iv) Before temporary removal of individual members begins, additional precautions shall be taken to ensure the safety of employees, such as installing other structural members to carry the loads imposed on the support system.

(v) Removal shall begin at, and progress from, the bottom of the excavation. Members shall be released slowly so as to note any indication of possible failure of the remaining members of the structure or possible cave-in of the sides of the excavation.

(vi) Backfilling shall progress together with the removal of support systems from excavations.

(2) Additional requirements for support systems for trench excavations.

(i) Excavation of material to a level no greater than 2 feet (.61 m) below the bottom of the members of a support system shall be permitted, but only if the system is designed to resist the forces calculated for the full depth of the trench, and there are no indications while the trench is open of a possible loss of soil from behind or below the bottom of the support system.

(ii) Installation of a support system shall be closely coordinated with the excavation of trenches.

(f) Sloping and benching systems. Employees shall not be permitted to work on the faces of sloped or benched excavations at levels above other employees except when employees at the lower levels are adequately protected from the hazard of falling, rolling, or sliding material or equipment.

(g) Shield systems.

(1) (i) Shield systems shall not be subjected to loads exceeding those which the system was designed to withstand.

(ii) Shields shall be installed in a manner to restrict lateral or other hazardous movement of the shield in the event of the application of sudden lateral loads.

(iii) Employees shall be protected from the hazard of cave-ins when entering or exiting the areas protected by shields.

(iv) Employees shall not be allowed in shields when shields are being installed, removed, or moved vertically.

(2) Additional requirement for shield systems used in trench excavations. Excavations of earth material to a level not greater than 2 feet (.61 m) below the bottom of a shield shall be permitted, but only if

the shield is designed to resist the forces calculated for the full depth of the trench, and there are no indications while the trench is open of a possible loss of soil from behind or below the bottom of the shield.

1926 SUBPART Q—CONCRETE AND MASONRY CONSTRUCTION

Fire Station 39's foundation was built with cast-in-place concrete and its superstructure is composed of masonry walls, concrete slab on metal decking, and structural steel. Chapter 6 discusses the foundation concrete and Chapter 7 examines masonry construction.

1926.700—Scope, Application, and Definitions Applicable To This Subpart

(a) Scope and application. This subpart sets forth requirements to protect all construction employees from the hazards associated with concrete and masonry construction operations performed in workplaces covered under 29 CFR Part 1926. In addition to the requirements in Subpart Q, other relevant provisions in Parts 1910 and 1926 apply to concrete and masonry construction operations.

1926.701—General Requirements

On Fire Station 39, vertical reinforcement was installed inside the masonry walls, creating the potential of impalement hazards.

(a) No construction loads shall be placed on a concrete structure or portion of a concrete structure unless the employer determines, based on information received from a person who is qualified in structural design, that the structure or portion of the structure is capable of supporting the loads.

(b) All protruding reinforcing steel, onto and into which employees could fall, shall be guarded to eliminate the hazard of impalement.

(c) Post-tensioning operations.

(d) No employee shall be permitted to ride concrete buckets.

(e) Working under loads.

  (1) No employee shall be permitted to work under concrete buckets while buckets are being elevated or lowered into position.

  (2) To the extent practical, elevated concrete buckets shall be routed so that no employee, or the fewest number of employees, are exposed to the hazards associated with falling concrete buckets.

(f) Personal protective equipment. No employee shall be permitted to apply a cement, sand, and water mixture through a pneumatic hose unless the employee is wearing protective head and face equipment.

1926.702—Requirements for Equipment and Tools

(a) Bulk cement storage.
   (1) Bulk storage bins, containers, and silos shall be equipped with the following:
      (i) Conical or tapered bottoms; and
      (ii) Mechanical or pneumatic means of starting the flow of material.
   (2) No employee shall be permitted to enter storage facilities unless the ejection system has been shut down, locked out, and tagged to indicate that the ejection system is not to be operated.

(b) Concrete mixers. Concrete mixers with one cubic yard (-8 m(3)) or larger loading skips shall be equipped with the following:
   (1) A mechanical device to clear the skip of materials; and
   (2) Guardrails installed on each side of the skip.

(c) Power concrete trowels. Powered and rotating type concrete troweling machines that are manually guided shall be equipped with a control switch that will automatically shut off the power whenever the hands of the operator are removed from the equipment handles.

(d) Concrete buggies. Concrete buggy handles shall not extend beyond the wheels on either side of the buggy.

(e) Concrete pumping systems.
   (1) Concrete pumping systems using discharge pipes shall be provided with pipe supports designed for 100 percent overload.
   (2) Compressed air hoses used on concrete pumping system shall be provided with positive fail-safe joint connectors to prevent separation of sections when pressurized.

(f) Concrete buckets.
   (1) Concrete buckets equipped with hydraulic or pneumatic gates shall have positive safety latches or similar safety devices installed to prevent premature or accidental dumping.
   (2) Concrete buckets shall be designed to prevent concrete from hanging up on top and the sides.

(g) Tremies. Sections of tremies and similar concrete conveyances shall be secured with wire rope (or equivalent materials) in addition to the regular couplings or connections.

(h) Bull floats. Bull float handles used where they might contact energized electrical conductors, shall be constructed of nonconductive material or insulated with a nonconductive sheath whose electrical and mechanical characteristics provide the equivalent protection of a handle constructed of nonconductive material.

(i) Masonry saws.
   (1) Masonry saw shall be guarded with a semicircular enclosure over the blade.
   (2) A method for retaining blade fragments shall be incorporated in the design of the semicircular enclosure.

(j) Lockout/Tagout procedures.

(1) No employee shall be permitted to perform maintenance or repair activity on equipment (such as compressors mixers, screens or pumps used for concrete and masonry construction activities) where the inadvertent operation of the equipment could occur and cause injury, unless all potentially hazardous energy sources have been locked out and tagged.

(2) Tags shall read Do Not Start or similar language to indicate that the equipment is not to be operated.

1926.703—Requirements for Cast-in-Place Concrete

(a) General requirements for formwork.

(1) Formwork shall be designed, fabricated, erected, supported, braced and maintained so that it will be capable of supporting without failure all vertical and lateral loads that may reasonably be anticipated to be applied to the formwork. Formwork which is designed, fabricated, erected, supported, braced and maintained in conformance with the Appendix to this section will be deemed to meet the requirements of this paragraph.

(2) Drawings or plans, including all revisions, for the jack layout, formwork (including shoring equipment), working decks, and scaffolds, shall be available at the jobsite.

(b) Shoring and reshoring.

(c) Vertical slip forms.

(d) Reinforcing steel.

(1) Reinforcing steel for walls, piers, columns, and similar vertical structures shall be adequately supported to prevent overturning and to prevent collapse.

(2) Employers shall take measures to prevent unrolled wire mesh from recoiling. Such measures may include, but are not limited to, securing each end of the roll or turning over the roll.

(e) Removal of formwork.

(1) Forms and shores (except those used for slabs on grade and slip forms) shall not be removed until the employer determines that the concrete has gained sufficient strength to support its weight and superimposed loads. Such determination shall be based on compliance with one of the following:

(i) The plans and specifications stipulate conditions for removal of forms and shores, and such conditions have been followed, or

(ii) The concrete has been properly tested with an appropriate ASTM standard test method designed to indicate the concrete compressive strength, and the test results indicate that the concrete has gained sufficient strength to support its weight and superimposed loads.

(2) Reshoring shall not be removed until the concrete being supported has attained adequate strength to support its weight and all loads in place upon it.

1926.706—Requirements for Masonry Construction

At the Fire Station 39 project, a significant amount of masonry wall was put in place around the apparatus bay area and called for the establishment of a limited access zone as discussed in Chapter 7.

(a) A limited access zone shall be established whenever a masonry wall is being constructed. The limited access zone shall conform to the following.

(1) The limited access zone shall be established prior to the start of construction of the wall.

(2) The limited access zone shall be equal to the height of the wall to [be] reconstructed plus four feet, and shall run the entire length of the wall.

(3) The limited access zone shall be established on the side of the wall which will be unscaffolded.

(4) The limited access zone shall be restricted to entry by employees actively engaged in constructing the wall. No other employees shall be permitted to enter the zone.

(5) The limited access zone shall remain in place until the wall is adequately supported to prevent overturning and to prevent collapse unless the height of [the] wall is over eight feet, in which case, the limited access zone shall remain in place until the requirements of paragraph (b) of this section have been met.

(b) All masonry walls over eight feet in height shall be adequately braced to prevent overturning and to prevent collapse unless the wall is adequately supported so that it will not overturn or collapse. The bracing shall remain in place until permanent supporting elements of the structure are in place.

1926 SUBPART R—STEEL ERECTION

The station house and the apparatus bay roof at the Fire Station 39 project were both supported by structural steel. Even though the general contractor most likely subcontracted the fabrication and erection of structural steel, the general contractor should still be aware of the general requirements in order to review the safety plan from the subcontractor and to make necessary field coordination.

1926.750—Scope

(a) This subpart sets forth requirements to protect employees from the hazards associated with steel erection activities involved in the construction, alteration, and/or repair of single and multi-story buildings, bridges, and other structures where steel erection occurs. The requirements of this subpart apply to employers engaged in steel erection unless otherwise specified. This subpart does not cover electrical transmission towers, communication and broadcast towers, or tanks.

(b) (1) Steel erection activities include hoisting, laying out, placing, connecting, welding, burning, guying, bracing, bolting, plumbing and rigging structural steel, steel joists and metal buildings; installing metal decking, curtain walls, window walls, siding systems, miscellaneous metals, ornamental iron and similar materials; and moving point-to-point while performing these activities.

(2) The following activities are covered by this subpart when they occur during and are a part of steel erection activities: rigging, hoisting, laying out, placing, connecting, guying, bracing, dismantling, burning, welding, bolting, grinding, sealing, caulking, and all related activities for construction, alteration and/or repair of materials and assemblies such as structural steel; ferrous metals and alloys; non-ferrous metals and alloys; glass; plastics and synthetic composite materials; structural metal framing and related bracing and assemblies; anchoring devices; structural cabling; cable stays; permanent and temporary bents and towers; falsework for temporary supports of permanent steel members; stone and other non-precast concrete architectural materials mounted on steel frames; safety systems for steel erection; steel and metal joists; metal decking and raceway systems and accessories; metal roofing and accessories; metal siding; bridge flooring; cold formed steel framing; elevator beams; grillage; shelf racks; multi-purpose supports; crane rails and accessories; miscellaneous, architectural and ornamental metals and metal work; ladders; railings; handrails; fences and gates; gratings; trench covers; floor plates; castings; sheet metal fabrications; metal panels and panel wall systems; louvers; column covers; enclosures and pockets; stairs; perforated metals; ornamental iron work, expansion control including bridge expansion joint assemblies; slide bearings; hydraulic structures; fascias; soffit panels; penthouse enclosures; skylights; joint fillers; gaskets; sealants and seals; doors; windows; hardware; detention/security equipment and doors, windows and hardware; conveying systems; building specialties; building equipment; machinery and plant equipment, furnishings and special construction.

1926.752—Site Layout, Site-Specific Erection Plan and Construction Sequence.

(a) Approval to begin steel erection. Before authorizing the commencement of steel erection, the controlling contractor shall ensure that the steel erector is provided with the following written notifications:

(1) The concrete in the footings, piers and walls and the mortar in the masonry piers and walls has attained, on the basis of an appropriate ASTM standard test method of field-cured samples, either 75 percent of the intended minimum compressive design strength or sufficient strength to support the loads imposed during steel erection.

(2) Any repairs, replacements and modifications to the anchor bolts were conducted in accordance with § 1926.755(b).

(b) Commencement of steel erection. A steel erection contractor shall not erect steel unless it has received written notification that the concrete in the footings, piers and walls or the mortar in the masonry piers and walls has attained, on the basis of an appropriate ASTM standard test method of field-cured samples, either 75 percent of the intended minimum compressive design strength or sufficient strength to support the loads imposed during steel erection.

(c) Site layout. The controlling contractor shall ensure that the following is provided and maintained:

(1) Adequate access roads into and through the site for the safe delivery and movement of derricks, cranes, trucks, other necessary equipment, and the material to be erected and means and methods for pedestrian and vehicular control. Exception: this requirement does not apply to roads outside of the construction site.

(2) A firm, properly graded, drained area, readily accessible to the work with adequate space for the safe storage of materials and the safe operation of the erector's equipment.

(d) Pre-planning of overhead hoisting operations. All hoisting operations in steel erection shall be pre-planned to ensure that the requirements of § 1926.753(d) are met.

(e) Site-specific erection plan. Where employers elect, due to conditions specific to the site, to develop alternate means and methods that provide employee protection in accordance with § 1926.753(c)(5), § 1926.757(a)(4) or § 1926.757(e)(4), a site-specific erection plan shall be developed by a qualified person and be available at the work site. Guidelines for establishing a site-specific erection plan are contained in Appendix A to this subpart.

1926.753—Hoisting and Rigging

Fire Station 39 is a structural steel building and a mobile crane was used on site to help hoist steel members into their respective locations. Chapter 7 discusses how this part of the standards impacts pretask planning and project scheduling.

(a) All the provisions of subpart CC apply to hoisting and rigging with the exception of § 1926.1431(a).

(b) In addition, paragraphs (c) through (e) of this section apply regarding the hazards associated with hoisting and rigging.

(c) (1) Pre-shift visual inspection of cranes.

(i) Cranes being used in steel erection activities shall be visually inspected prior to each shift by a competent person; the inspection shall include observation for deficiencies during operation. At a minimum this inspection shall include the following:

(A) All control mechanisms for maladjustments;

(B) Control and drive mechanism for excessive wear of components and contamination by lubricants, water or other foreign matter;

(C) Safety devices, including but not limited to boom angle indicators, boom stops, boom kick out devices, anti-two block devices, and load moment indicators where required;

(D) Air, hydraulic, and other pressurized lines for deterioration or leakage, particularly those which flex in normal operation;

(E) Hooks and latches for deformation, chemical damage, cracks, or wear;

(F) Wire rope reeving for compliance with hoisting equipment manufacturer's specifications;

(G) Electrical apparatus for malfunctioning, signs of excessive deterioration, dirt, or moisture accumulation;

(H) Hydraulic system for proper fluid level;

(I) Tires for proper inflation and condition;

(J) Ground conditions around the hoisting equipment for proper support, including ground settling under and around outriggers, ground water accumulation, or similar conditions;

(K) The hoisting equipment for level position; and

(L) The hoisting equipment for level position after each move and setup.

(ii) If any deficiency is identified, an immediate determination shall be made by the competent person as to whether the deficiency constitutes a hazard.

(iii) If the deficiency is determined to constitute a hazard, the hoisting equipment shall be removed from service until the deficiency has been corrected.

(iv) The operator shall be responsible for those operations under the operator's direct control. Whenever there is any doubt as to safety, the operator shall have the authority to stop and refuse to handle loads until safety has been assured.

(2) A qualified rigger (a rigger who is also a qualified person) shall inspect the rigging prior to each shift in accordance with § 1926.251.

(3) The headache ball, hook or load shall not be used to transport personnel except as provided in paragraph (c)(4) of this section.

(4) Cranes or derricks may be used to hoist employees on a personnel platform when work under this subpart is being conducted, provided that all provisions of § 1926.1431 (except for § 1926.1431(a)) are met.

(5) Safety latches on hooks shall not be deactivated or made inoperable except:

(i) When a qualified rigger has determined that the hoisting and placing of purlins and single joists can be performed more safely by doing so; or

(ii) When equivalent protection is provided in a site-specific erection plan.

(d) Working under loads.

(1) Routes for suspended loads shall be pre-planned to ensure that no employee is required to work directly below a suspended load except for:

(i) Employees engaged in the initial connection of the steel; or

(ii) Employees necessary for the hooking or unhooking of the load.

(2) When working under suspended loads, the following criteria shall be met:
   (i) Materials being hoisted shall be rigged to prevent unintentional displacement;
   (ii) Hooks with self-closing safety latches or their equivalent shall be used to prevent components from slipping out of the hook; and
   (iii) All loads shall be rigged by a qualified rigger.

(e) Multiple lift rigging procedure.
   (1) A multiple lift shall only be performed if the following criteria are met:
      (i) A multiple lift rigging assembly is used;
      (ii) A maximum of five members are hoisted per lift;
      (iii) Only beams and similar structural members are lifted; and
      (iv) All employees engaged in the multiple lift have been trained in these procedures in accordance with § 1926.761(c)(1).
      (v) No crane is permitted to be used for a multiple lift where such use is contrary to the manufacturer's specifications and limitations.
   (2) Components of the multiple lift rigging assembly shall be specifically designed and assembled with a maximum capacity for total assembly and for each individual attachment point. This capacity, certified by the manufacturer or a qualified rigger, shall be based on the manufacturer's specifications with a 5 to 1 safety factor for all components.
   (3) The total load shall not exceed:
      (i) The rated capacity of the hoisting equipment specified in the hoisting equipment load charts;
      (ii) The rigging capacity specified in the rigging rating chart.
   (4) The multiple lift rigging assembly shall be rigged with members:
      (i) Attached at their center of gravity and maintained reasonably level;
      (ii) Rigged from top down; and
      (iii) Rigged at least 7 feet (2.1 m) apart.
   (5) The members on the multiple lift rigging assembly shall be set from the bottom up.
   (6) Controlled load lowering shall be used whenever the load is over the connectors.

1926.754—Structural Steel Assembly

(a) Structural stability shall be maintained at all times during the erection process.

(b) The following additional requirements shall apply for multi-story structures:
   (1) The permanent floors shall be installed as the erection of structural members progresses, and there shall be not more than eight stories between the erection floor and the upper-most permanent floor, except where the structural integrity is maintained as a result of the design.
   (2) At no time shall there be more than four floors or 48 feet (14.6 m), whichever is less, of unfinished bolting or welding above the foundation or uppermost permanently secured floor, except where the structural integrity is maintained as a result of the design.

(3) A fully planked or decked floor or nets shall be maintained within two stories or 30 feet (9.1 m), whichever is less, directly under any erection work being performed.

(c) Walking/working surfaces—shear connectors and other similar devices.

(1) Tripping hazards. Shear connectors (such as headed steel studs, steel bars or steel lugs), reinforcing bars, deformed anchors or threaded studs shall not be attached to the top flanges of beams, joists or beam attachments so that they project vertically from or horizontally across the top flange of the member until after the metal decking, or other walking/working surface, has been installed.

(2) Installation of shear connectors on composite floors, roofs and bridge decks. When shear connectors are used in construction of composite floors, roofs and bridge decks, employees shall lay out and install the shear connectors after the metal decking has been installed, using the metal decking as a working platform. Shear connector shall not be installed from within a controlled decking zone (CDZ), as specified in § 1926.760(c)(8).

(d) Plumbing-up.

(1) When deemed necessary by a competent person, plumbing-up equipment shall be installed in conjunction with the steel erection process to ensure the stability of the structure.

(2) When used, plumbing-up equipment shall be in place and properly installed before the structure is loaded with construction material such as loads of joists, bundles of decking or bundles of bridging.

(3) Plumbing-up equipment shall be removed only with the approval of a competent person.

(e) Metal decking.

(1) Hoisting, landing and placing of metal decking bundles.

(i) Bundle packaging and strapping shall not be used for hoisting unless specifically designed for that purpose.

(ii) If loose items such as dunnage, flashing, or other materials are placed on the top of metal decking bundles to be hoisted, such items shall be secured to the bundles.

(iii) Bundles of metal decking on joists shall be landed in accordance with § 1926.757(e)(4).

(iv) Metal decking bundles shall be landed on framing members so that enough support is provided to allow the bundles to be unbanded without dislodging the bundles from the supports.

(v) At the end of the shift or when environmental or jobsite conditions require, metal decking shall be secured against displacement.

(2) Roof and floor holes and openings. Metal decking at roof and floor holes and openings shall be installed as follows:

(i) Framed metal deck openings shall have structural members turned down to allow continuous deck installation except where not allowed by structural design constraints or constructability.

(ii) Roof and floor holes and openings shall be decked over. Where large size, configuration or other structural design does not allow openings to be decked over (such as elevator shafts, stair wells, etc.) employees shall be protected in accordance with § 1926.760(a)(1).

(iii) Metal decking holes and openings shall not be cut until immediately prior to being permanently filled with the equipment or structure needed or intended to fulfill its specific use and which meets the strength requirements of paragraph (e)(3) of this section, or shall be immediately covered.

(3) Covering roof and floor openings.

(i) Covers for roof and floor openings shall be capable of supporting, without failure, twice the weight of the employees, equipment and materials that may be imposed on the cover at any one time.

(ii) All covers shall be secured when installed to prevent accidental displacement by the wind, equipment or employees.

(iii) All covers shall be painted with high-visibility paint or shall be marked with the word "HOLE" or "COVER" to provide warning of the hazard.

(iv) Smoke dome or skylight fixtures that have been installed, are not considered covers for the purpose of this section unless they meet the strength requirements of paragraph (e)(3)(i) of this section.

(4) Decking gaps around columns. Wire mesh, exterior plywood, or equivalent, shall be installed around columns where planks or metal decking do not fit tightly. The materials used must be of sufficient strength to provide fall protection for personnel and prevent objects from falling through.

(5) Installation of metal decking.

(i) Except as provided in § 1926.760(c), metal decking shall be laid tightly and immediately secured upon placement to prevent accidental movement or displacement.

(ii) During initial placement, metal decking panels shall be placed to ensure full support by structural members.

(6) Derrick floors.

(i) A derrick floor shall be fully decked and/or planked and the steel member connections completed to support the intended floor loading.

(ii) Temporary loads placed on a derrick floor shall be distributed over the underlying support members so as to prevent local overloading of the deck material.

1926.755—Column Anchorage

(a) General requirements for erection stability.

(1) All columns shall be anchored by a minimum of 4 anchor rods (anchor bolts).

(2) Each column anchor rod (anchor bolt) assembly, including the column-to-base plate weld and the column foundation, shall be designed to resist a minimum eccentric gravity load of 300 pounds (136.2 kg) located 18 inches (.46 m) from the extreme outer face of the column in each direction at the top of the column shaft.

(3) Columns shall be set on level finished floors, pre-grouted leveling plates, leveling nuts, or shim packs which are adequate to transfer the construction loads.

(4) All columns shall be evaluated by a competent person to determine whether guying or bracing is needed; if guying or bracing is needed, it shall be installed.

(b) Repair, replacement or field modification of anchor rods (anchor bolts).

(1) Anchor rods (anchor bolts) shall not be repaired, replaced or field-modified without the approval of the project structural engineer of record.

(2) Prior to the erection of a column, the controlling contractor shall provide written notification to the steel erector if there has been any repair, replacement or modification of the anchor rods (anchor bolts) of that column.

1926.756—Beams and Columns

(a) General.

(1) During the final placing of solid web structural members, the load shall not be released from the hoisting line until the members are secured with at least two bolts per connection, of the same size and strength as shown in the erection drawings, drawn up wrench-tight or the equivalent as specified by the project structural engineer of record, except as specified in paragraph (b) of this section.

(2) A competent person shall determine if more than two bolts are necessary to ensure the stability of cantilevered members; if additional bolts are needed, they shall be installed.

(b) Diagonal bracing. Solid web structural members used as diagonal bracing shall be secured by at least one bolt per connection drawn up wrench-tight or the equivalent as specified by the project structural engineer of record.

(c) (1) Double connections at columns and/or at beam webs over a column. When two structural members on opposite sides of a column web, or a beam web over a column, are connected sharing common connection holes, at least one bolt with its wrench-tight nut shall remain connected to the first member unless a shop-attached or field-attached seat or equivalent connection device is supplied with the member to secure the first member and prevent the column from being displaced (See Appendix H to this subpart for examples of equivalent connection devices).

(2) If a seat or equivalent device is used, the seat (or device) shall be designed to support the load during the double connection process. It shall be adequately bolted or welded to both a supporting member and the first member before the nuts on the shared bolts are removed to make the double connection.

(d) Column splices. Each column splice shall be designed to resist a minimum eccentric gravity load of 300 pounds (136.2 kg) located 18 inches (.46 m) from the extreme outer face of the column in each direction at the top of the column shaft.

(e) Perimeter columns. Perimeter columns shall not be erected unless:

(1) The perimeter columns extend a minimum of 48 inches (1.2 m) above the finished floor to permit installation of perimeter safety cables prior to erection of the next tier, except where constructability does not allow (see Appendix F to this subpart);

(2) The perimeter columns have holes or other devices in or attached to perimeter columns at 42–45 inches (107–114 cm) above the finished floor and the midpoint between the finished floor and the top cable to permit installation of perimeter safety cables required by § 1926.760(a)(2), except where constructability does not allow. (See Appendix F to this subpart).

1926.759—Falling Object Protection

(a) Securing loose items aloft. All materials, equipment, and tools, which are not in use while aloft, shall be secured against accidental displacement.

(b) Protection from falling objects other than materials being hoisted. The controlling contractor shall bar other construction processes below steel erection unless overhead protection for the employees below is provided.

1926.760—Fall Protection

Chapter 7 discusses the working elevation for the steel erection activities at the Fire Station 39 project and examines how to apply the fall protection requirement from this part of the OSHA standard considering the site conditions. Chapter 10 further illustrates the practical considerations when evaluating different fall protection options based on the site conditions.

(a) (1) Except as provided by paragraph (a)(3) of this section, each employee engaged in a steel erection activity who is on a walking/working surface with an unprotected side or edge more than 15 feet (4.6 m) above a lower level shall be protected from fall hazards by guardrail systems, safety net systems, personal fall arrest systems, positioning device systems or fall restraint systems.

(2) On multi-story structures, perimeter safety cables shall be installed at the final interior and exterior perimeters of the floors as soon as the metal decking has been installed.

(3) Connectors and employees working in controlled decking zones shall be protected from fall hazards as provided in paragraphs (b) and (c) of this section, respectively.

(b) Connectors. Each connector shall:

(1) Be protected in accordance with paragraph (a)(1) of this section from fall hazards of more than two stories or 30 feet (9.1 m) above a lower level, whichever is less;

(2) Have completed connector training in accordance with § 1926.761; and

(3) Be provided, at heights over 15 and up to 30 feet above a lower level, with a personal fall arrest system, positioning device system or fall restraint system and wear the equipment necessary to be able to be tied off; or be provided with other means of protection from fall hazards in accordance with paragraph (a)(1) of this section.

(c) Controlled Decking Zone (CDZ). A controlled decking zone may be established in that area of the structure over 15 and up to 30 feet above a lower level where metal decking is initially being installed and forms the leading edge of a work area. In each CDZ, the following shall apply:
(1) Each employee working at the leading edge in a CDZ shall be protected from fall hazards of more than two stories or 30 feet (9.1 m), whichever is less.
(2) Access to a CDZ shall be limited to only those employees engaged in leading edge work.
(3) The boundaries of a CDZ shall be designated and clearly marked. The CDZ shall not be more than 90 feet (27.4 m) wide and 90 (27.4 m) feet deep from any leading edge. The CDZ shall be marked by the use of control lines or the equivalent. Examples of acceptable procedures for demarcating CDZ's can be found in Appendix D to this subpart.
(4) Each employee working in a CDZ shall have completed CDZ training in accordance with § 1926.761.
(5) Unsecured decking in a CDZ shall not exceed 3,000 square feet (914.4 m^2).
(6) Safety deck attachments shall be performed in the CDZ from the leading edge back to the control line and shall have at least two attachments for each metal decking panel.
(7) Final deck attachments and installation of shear connectors shall not be performed in the CDZ.

(d) Criteria for fall protection equipment.
(1) Guardrail systems, safety net systems, personal fall arrest systems, positioning device systems and their components shall conform to the criteria in § 1926.502 (see Appendix G to this subpart).
(2) Fall arrest system components shall be used in fall restraint systems and shall conform to the criteria in § 1926.502 (see Appendix G). Either body belts or body harnesses shall be used in fall restraint systems.
(3) Perimeter safety cables shall meet the criteria for guardrail systems in § 1926.502 (see Appendix G).

(e) Custody of fall protection. Fall protection provided by the steel erector shall remain in the area where steel erection activity has been completed, to be used by other trades, only if the controlling contractor or its authorized representative:
(1) Has directed the steel erector to leave the fall protection in place; and
(2) Has inspected and accepted control and responsibility of the fall protection prior to authorizing persons other than steel erectors to work in the area.

1926 SUBPART X—LADDERS

Ladders are common means to access higher working elevations and were heavily used on the Fire Station 39 project, especially during exterior enclosure and interior construction. Proper ladder uses citing parts of the OSHA standards Subpart X are discussed in detail in Chapter 9.

1926.1050—Scope, Application, and Definitions Applicable To This Subpart

(a) Scope and application. This subpart applies to all stairways and ladders used in construction, alteration, repair (including painting and decorating), and demolition workplaces covered under 29 CFR part 1926, and also sets forth, in specified circumstances, when ladders and stairways are required to be provided. Additional requirements for ladders used on or with scaffolds are contained in subpart L—Scaffolds. This subpart does not apply to integral components of equipment covered by subpart CC. Subpart CC exclusively sets forth the circumstances when ladders and stairways must be provided on equipment covered by subpart CC.

1926.1051—General Requirements

(a) A stairway or ladder shall be provided at all personnel points of access where there is a break in elevation of 19 inches (48 cm) or more, and no ramp, runway, sloped embankment, or personnel hoist is provided.
   (1) Employees shall not use any spiral stairways that will not be a permanent part of the structure on which construction work is being performed.
   (2) A double-cleated ladder or two or more separate ladders shall be provided when ladders are the only mean of access or exit from a working area for 25 or more employees, or when a ladder is to serve simultaneous two-way traffic.
   (3) When a building or structure has only one point of access between levels, that point of access shall be kept clear to permit free passage of employees. When work must be performed or equipment must be used such that free passage at that point of access is restricted, a second point of access shall be provided and used.
   (4) When a building or structure has two or more points of access between levels, at least one point of access shall be kept clear to permit free passage of employees.

(b) Employers shall provide and install all stairway and ladder fall protection systems required by this subpart and shall comply with all other pertinent requirements of this subpart before employees begin the work that necessitates the installation and use of stairways, ladders, and their respective fall protection systems.

1926.1052—Stairways

(a) General. The following requirements apply to all stairways as indicated:
   (1) Stairways that will not be a permanent part of the structure on which construction work is being performed shall have landings of not less than 30 inches (76 cm) in the direction of travel and extend at least 22 inches (56 cm) in width at every 12 feet (3.7 m) or less of vertical rise.
   (2) Stairs shall be installed between 30 deg. and 50 deg. from horizontal.

(3) Riser height and tread depth shall be uniform within each flight of stairs, including any foundation structure used as one or more treads of the stairs. Variations in riser height or tread depth shall not be over ¼-inch (0.6 cm) in any stairway system.

(4) Where doors or gates open directly on a stairway, a platform shall be provided, and the swing of the door shall not reduce the effective width of the platform to less than 20 inches (51 cm).

(5) Metal pan landings and metal pan treads, when used, shall be secured in place before filling with concrete or other material.

(6) All parts of stairways shall be free of hazardous projections, such as protruding nails.

(7) Slippery conditions on stairways shall be eliminated before the stairways are used to reach other levels.

(b) Temporary service. The following requirements apply to all stairways as indicated:

(1) Except during stairway construction, foot traffic is prohibited on stairways with pan stairs where the treads and/or landings are to be filled in with concrete or other material at a later date, unless the stairs are temporarily fitted with wood or other solid material at least to the top edge of each pan. Such temporary treads and landings shall be replaced when worn below the level of the top edge of the pan.

(2) Except during stairway construction, foot traffic is prohibited on skeleton metal stairs where permanent treads and/or landings are to be installed at a later date, unless the stairs are fitted with secured temporary treads and landings long enough to cover the entire tread and/or landing area.

(3) Treads for temporary service shall be made of wood or other solid material, and shall be installed the full width and depth of the stair.

(c) Stairrails and handrails. The following requirements apply to all stairways as indicated:

(1) Stairways having four or more risers or rising more than 30 inches (76 cm), whichever is less, shall be equipped with:

(i) At least one handrail; and

(ii) One stairrail system along each unprotected side or edge.

Note: When the top edge of a stairrail system also serves as a handrail, paragraph (c)(7) of this section applies.

(2) Winding and spiral stairways shall be equipped with a handrail offset sufficiently to prevent walking on those portions of the stairways where the tread width is less than 6 inches (15 cm).

(3) The height of stairrails shall be as follows:

(i) Stairrails installed after March 15, 1991, shall be not less than 36 inches (91.5 cm) from the upper surface of the stairrail system to the surface of the tread, in line with the face of the riser at the forward edge of the tread.

(ii) Stairrails installed before March 15, 1991, shall be not less than 30 inches (76 cm) nor more than 34 inches (86 cm) from the upper surface of the stairrail system to the surface of the tread, in line with the face of the riser at the forward edge of the tread.

(4) Midrails, screens, mesh, intermediate vertical members, or equivalent intermediate structural members, shall be provided between the top rail of the stairrail system and the stairway steps.

(i) Midrails, when used, shall be located at a height midway between the top edge of the stairrail system and the stairway steps.

(ii) Screens or mesh, when used, shall extend from the top rail to the stairway step, and along the entire opening between top rail supports.

(iii) When intermediate vertical members, such as balusters, are used between posts, they shall be not more than 19 inches (48 cm) apart.

(iv) Other structural members, when used, shall be installed such that there are no openings in the stairrail system that are more than 19 inches (48 cm) wide.

(5) Handrails and the top rails of stairrail systems shall be capable of withstanding, without failure, a force of at least 200 pounds (890 N) applied within 2 inches (5 cm) of the top edge, in any downward or outward direction, at any point along the top edge.

(6) The height of handrails shall be not more than 37 inches (94 cm) nor less than 30 inches (76 cm) from the upper surface of the handrail to the surface of the tread, in line with the face of the riser at the forward edge of the tread.

(7) When the top edge of a stairrail system also serves as a handrail, the height of the top edge shall be not more than 37 inches (94 cm) nor less than 36 inches (91.5 cm) from the upper surface of the stairrail system to the surface of the tread, in line with the face of the riser at the forward edge of the tread.

(8) Stairrail systems and handrails shall be so surfaced as to prevent injury to employees from punctures or lacerations, and to prevent snagging of clothing.

(9) Handrails shall provide an adequate handhold for employees grasping them to avoid falling.

(10) The ends of stairrail systems and handrails shall be constructed so as not to constitute a projection hazard.

(11) Handrails that will not be a permanent part of the structure being built shall have a minimum clearance of 3 inches (8 cm) between the handrail and walls, stairrail systems, and other objects.

(12) Unprotected sides and edges of stairway landings shall be provided with guardrail systems. Guardrail system criteria are contained in subpart M of this part.

1926.1053—Ladders

(a) General. The following requirements apply to all ladders as indicated, including job-made ladders.

(1) Ladders shall be capable of supporting the following loads without failure:

(i) Each self-supporting portable ladder: At least four times the maximum intended load, except that each extra-heavy-duty type 1A metal or plastic ladder shall sustain at least 3.3 time[s] the maximum intended load. The ability of a ladder to sustain the loads indicated in this paragraph shall be determined by applying or transmitting the requisite load to the ladder in a downward vertical direction. Ladders built and tested in conformance with the applicable provisions of appendix A of this subpart will be deemed to meet this requirement.

(ii) Each portable ladder that is not self-supporting: At least four times the maximum intended load, except that each extra-heavy-duty type 1A metal or plastic ladders shall sustain at least 3.3 times the maximum intended load. The ability of a ladder to sustain the loads indicated in this paragraph shall be determined by applying or transmitting the requisite load to the ladder in a downward vertical direction when the ladder is placed at an angle of 75 ½ degrees from the horizontal. Ladders built and tested in conformance with the applicable provisions of appendix A will be deemed to meet this requirement.

(iii) Each fixed ladder: At least two loads of 250 pounds (114 kg) each, concentrated between any two consecutive attachments (the number and position of additional concentrated loads of 250 pounds (114 kg) each, determined from anticipated usage of the ladder, shall also be included), plus anticipated loads caused by ice buildup, winds, rigging, and impact loads resulting from the use of ladder safety devices. Each step or rung shall be capable of supporting a single concentrated load of a least 250 pounds (114 kg) applied in the middle of the step or rung. Ladders built in conformance with the applicable provisions of appendix A will be deemed to meet this requirement.

(2) Ladder rungs, cleats, and steps shall be parallel, level, and uniformly spaced when the ladder is in position for use.

(i) Rungs, cleats, and steps of portable ladders (except as provided below) and fixed ladders (including individual-rung/step ladders) shall be spaced not less than 10 inches (25 cm) apart, nor more than 14 inches (36 cm) apart, as measured between center lines of the rungs, cleats and steps.

(ii) Rungs, cleats, and steps of step stools shall be not less than 8 inches (20 cm) apart, nor more than 12 inches (31 cm) apart, as measured between center lines of the rungs, cleats, and steps.

(iii) Rungs, cleats, and steps of the base section of extension trestle ladders shall be not less than 8 inches (20 cm) nor more than 18 inches (46 cm) apart, as measured between center lines of the rungs, cleats, and steps. The rung spacing on the extension section of the extension trestle ladder shall be not less than 6 inches (15 cm) nor more than 12 inches (31 cm), as measured between center lines of the rungs, cleats, and steps.

(4) (i) The minimum clear distance between the sides of individual-rung/step ladders and the minimum clear distance between the side rails of other fixed ladders shall be 16 inches (41 cm).

(ii) The minimum clear distance between side rails for all portable ladders shall be 11½ inches (29 cm).

(5) The rungs of individual-rung/step ladders shall be shaped such that employees' feet cannot slide off the end of the rungs.

(6) (i) The rungs and steps of fixed metal ladders manufactured after March 15, 1991, shall be corrugated, knurled, dimpled, coated with skid-resistant material, or otherwise treated to minimize slipping.

(ii) The rungs and steps of portable metal ladders shall be corrugated, knurled, dimpled, coated with skid-resistant material, or otherwise treated to minimize slipping.

(7) Ladders shall not be tied or fastened together to provide longer sections unless they are specifically designed for such use.

(8) A metal spreader or locking device shall be provided on each stepladder to hold the front and back sections in an open position when the ladder is being used.

(9) When splicing is required to obtain a given length of side rail, the resulting side rail must be at least equivalent in strength to a one-piece side rail made of the same material.

(10) Except when portable ladders are used to gain access to fixed ladders (such as those on utility towers, billboards, and other structures where the bottom of the fixed ladder is elevated to limit access), when two or more separate ladders are used to reach an elevated work area, the ladders shall be offset with a platform or landing between the ladders. (The requirements to have guardrail systems with toeboards for falling object and overhead protection on platforms and landings are set forth in subpart M of this part.)

(11) Ladder components shall be surfaced so as to prevent injury to an employee from punctures or lacerations, and to prevent snagging of clothing.

(12) Wood ladders shall not be coated with any opaque covering, except for identification or warning labels which may be placed on one face only of a side rail.

(13) The minimum perpendicular clearance between fixed ladder rungs, cleats, and steps, and any obstruction behind the ladder shall be 7 inches (18 cm), except in the case of an elevator pit ladder for which a minimum perpendicular clearance of 4½ inches (11 cm) is required.

(14) The minimum perpendicular clearance between the center line of fixed ladder rungs, cleats, and steps, and any obstruction on the climbing side of the ladder shall be 30 inches (76 cm), except as provided in paragraph (a)(15) of this section.

(15) When unavoidable obstructions are encountered, the minimum perpendicular clearance between the centerline of fixed ladder rungs, cleats, and steps, and the obstruction on the climbing side of the ladder may be reduced to 24 inches (61 cm), provided that a deflection device is installed to guide employees around the obstruction.

(16) Through fixed ladders at their point of access/egress shall have a step-across distance of not less than 7 inches (18 cm) nor more than 12 inches (30 cm) as measured from the centerline of the steps or rungs to the nearest edge of the landing area. If the normal step-across distance exceeds 12 inches (30 cm), a landing platform shall be provided to reduce the distance to the specified limit.

(17) Fixed ladders without cages or wells shall have a clear width to the nearest permanent object of at least 15 inches (30 cm) on each side of the centerline of the ladder.

(18) Fixed ladders shall be provided with cages, wells, ladder safety devices, or self-retracting lifelines where the length of climb is less than 24 feet (7.3 m) but the top of the ladder is at a distance greater than 24 feet (7.3 m) above lower levels.

(19) Where the total length of a climb equals or exceeds 24 feet (7.3 m), fixed ladders shall be equipped with one of the following:

(i) Ladder safety devices; or

(ii) Self-retracting lifelines, and rest platforms at intervals not to exceed 150 feet (45.7 m); or

(iii) A cage or well, and multiple ladder sections, each ladder section not to exceed 50 feet (15.2 m) in length. Ladder sections shall be offset from adjacent sections, and landing platforms shall be provided at maximum intervals of 50 feet (15.2 m).

(20) Cages for fixed ladders shall conform to all of the following:

(i) Horizontal bands shall be fastened to the side rails of rail ladders, or directly to the structure, building, or equipment for individual-rung ladders;

(ii) Vertical bars shall be on the inside of the horizontal bands and shall be fastened to them;

(iii) Cages shall extend not less than 27 inches (66 cm), or more than 30 inches (76 cm) from the centerline of the step or rung (excluding the flare at the bottom of the cage), and shall not be less than 27 inches (68 cm) in width;

(iv) The inside of the cage shall be clear of projections;

(v) Horizontal bands shall be spaced not more than 4 feet (1.2 m) on center vertically;

(vi) Vertical bars shall be spaced at intervals not more than 9½ inches (24 cm) on center horizontally;

(vii) the bottom of the cage shall be at a level not less than 7 feet (2.1 m) nor more than 8 feet (2.4 m) above the point of access to the bottom of the ladder. The bottom of the cage shall be flared not less than 4 inches (10 cm) all around within the distance between the bottom horizontal band and the next higher band;

(viii) The top of the cage shall be a minimum of 42 inches (1.1 m) above the top of the platform, or the point of access at the top of the ladder, with provision for access to the platform or other point of access.

(21) Wells for fixed ladders shall conform to all of the following:
  (i) They shall completely encircle the ladder;
  (ii) They shall be free of projections;
  (iii) Their inside face on the climbing side of the ladder shall extend not less than 27 inches (68 cm) nor more than 30 inches (76 cm) from the centerline of the step or rung;
  (iv) The inside clear width shall be at least 30 inches (76 cm);
  (v) The bottom of the wall on the access side shall start at a level not less than 7 feet (2.1 m) nor more than 8 feet (2.4 m) above the point of access to the bottom of the ladder.

(22) Ladder safety devices, and related support systems, for fixed ladders shall conform to all of the following:
  (i) They shall be capable of withstanding without failure a drop test consisting of an 18-inch (41 cm) drop of a 500-pound (226 kg) weight;
  (ii) They shall permit the employee using the device to ascend or descend without continually having to hold, push, or pull any part of the device, leaving both hands free for climbing;
  (iii) They shall be activated within 2 feet (.61 m) after a fall occurs, and limit the descending velocity of an employee to 7 feet/sec. (2.1 m/sec.) or less;
  (iv) The connection between the carrier or lifeline and the point of attachment to the body belt or harness shall not exceed 9 inches (23 cm) in length.

(23) The mounting of ladder safety devices for fixed ladders shall conform to the following:
  (i) Mountings for rigid carriers shall be attached at each end of the carrier, with intermediate mountings, as necessary, spaced along the entire length of the carrier, to provide the strength necessary to stop employees' falls;
  (ii) Mountings for flexible carriers shall be attached at each end of the carrier. When the system is exposed to wind, cable guides for flexible carriers shall be installed at a minimum spacing of 25 feet (7.6 m) and maximum spacing of 40 feet (12.2 m) along the entire length of the carrier, to prevent wind damage to the system.
  (iii) The design and installation of mountings and cable guides shall not reduce the design strength of the ladder.

(24) The side rails of through or side-step fixed ladders shall extend 42 inches (1.1 m) above the top of the access level or landing platform served by the ladder. For a parapet ladder, the access level shall be the roof if the parapet is cut to permit passage through the parapet; if the parapet is continuous, the access level shall be the top of the parapet.

(25) For through-fixed-ladder extensions, the steps or rungs shall be omitted from the extension and the extension of the side rails shall be flared to provide not less than 24 inches (61 cm) nor more than 30 inches (76 cm) clearance between side rails. Where ladder safety devices are provided,

the maximum clearance between side rails of the extensions shall not exceed 36 inches (91 cm).

(26) For side-step fixed ladders, the side rails and the steps or rungs shall be continuous in the extension.

(27) Individual-rung/step ladders, except those used where their access openings are covered with manhole covers or hatches, shall extend at least 42 inches (1.1 m) above an access level or landing platform either by the continuation of the rung spacings as horizontal grab bars or by providing vertical grab bars that shall have the same lateral spacing as the vertical legs of the rungs.

(b) Use. The following requirements apply to the use of all ladders, including job-made ladders, except as otherwise indicated:

(1) When portable ladders are used for access to an upper landing surface, the ladder side rails shall extend at least 3 feet (.9 m) above the upper landing surface to which the ladder is used to gain access; or, when such an extension is not possible because of the ladder's length, then the ladder shall be secured at its top to a rigid support that will not deflect, and a grasping device, such as a grabrail, shall be provided to assist employees in mounting and dismounting the ladder. In no case shall the extension be such that ladder deflection under a load would, by itself, cause the ladder to slip off its support.

(2) Ladders shall be maintained free of oil, grease, and other slipping hazards.

(3) Ladders shall not be loaded beyond the maximum intended load for which they were built, nor beyond their manufacturer's rated capacity.

(4) Ladders shall be used only for the purpose for which they were designed.

(5) (i) Non-self-supporting ladders shall be used at an angle such that the horizontal distance from the top support to the foot of the ladder is approximately one-quarter of the working length of the ladder (the distance along the ladder between the foot and the top support).

(ii) Wood job-made ladders with spliced side rails shall be used at an angle such that the horizontal distance is one-eighth the working length of the ladder.

(iii) Fixed ladders shall be used at a pitch no greater than 90 degrees from the horizontal, as measured to the back side of the ladder.

(6) Ladders shall be used only on stable and level surfaces unless secured to prevent accidental displacement.

(7) Ladders shall not be used on slippery surfaces unless secured or provided with slip-resistant feet to prevent accidental displacement. Slip-resistant feet shall not be used as a substitute for care in placing, lashing, or holding a ladder that is used upon slippery surfaces including, but not limited to, flat metal or concrete surfaces that are constructed so they cannot be prevented from becoming slippery.

(8) Ladders placed in any location where they can be displaced by workplace activities or traffic, such as in passageways, doorways, or driveways, shall be secured to prevent accidental displacement, or a barricade shall be used to keep the activities or traffic away from the ladder.

(9) The area around the top and bottom of ladders shall be kept clear.
(10) The top of a non-self-supporting ladder shall be placed with the two rails supported equally unless it is equipped with a single support attachment.
(11) Ladders shall not be moved, shifted, or extended while occupied.
(12) Ladders shall have nonconductive siderails if they are used where the employee or the ladder could contact exposed energized electrical equipment, except as provided in 1926.951(c)(1) of this part.
(13) The top or top step of a stepladder shall not be used as a step.
(14) Cross-bracing on the rear section of stepladders shall not be used for climbing unless the ladders are designed and provided with steps for climbing on both front and rear sections.
(15) Ladders shall be inspected by a competent person for visible defects on a periodic basis and after any occurrence that could affect their safe use.
(16) Portable ladders with structural defects, such as, but not limited to, broken or missing rungs, cleats, or steps, broken or split rails, corroded components, or other faulty or defective components, shall either be immediately marked in a manner that readily identifies them as defective, or be tagged with "Do Not Use" or similar language, and shall be withdrawn from service until repaired.
(17) Fixed ladders with structural defects, such as, but not limited to, broken or missing rungs, cleats, or steps, broken or split rails, or corroded components, shall be withdrawn from service until repaired. The requirement to withdraw a defective ladder from service is satisfied if the ladder is either:
    (i) Immediately tagged with "Do Not Use" or similar language;
    (ii) Marked in a manner that readily identifies it as defective;
    (iii) Or blocked (such as with a plywood attachment that spans several rungs).
(18) Ladder repairs shall restore the ladder to a condition meeting its original design criteria, before the ladder is returned to use.
(19) Single-rail ladders shall not be used.
(20) When ascending or descending a ladder, the user shall face the ladder.
(21) Each employee shall use at least one hand to grasp the ladder when progressing up and/or down the ladder.
(22) An employee shall not carry any object or load that could cause the employee to lose balance and fall.

1926 SUBPART CC—CRANES AND DERRICKS IN CONSTRUCTION

A mobile crane was used for steel construction at the Fire Station 39 project and its related safety precautions referencing the OSHA Subpart CC standards are discussed in Chapter 7, Safety for the Superstructure. OSHA made significant changes to its crane-related standards with the final version being published in 2010. In particular, rules on the assessment of ground conditions and the procedures for working in the vicinity of power lines are included.

1926.1400—Scope

(a) This standard applies to power-operated equipment, when used in construction, that can hoist, lower and horizontally move a suspended load. Such equipment includes, but is not limited to: Articulating cranes (such as knuckle-boom cranes); crawler cranes; floating cranes; cranes on barges; locomotive cranes; mobile cranes (such as wheel-mounted, rough-terrain, all-terrain, commercial truck-mounted, and boom truck cranes); multi-purpose machines when configured to hoist and lower (by means of a winch or hook) and horizontally move a suspended load; industrial cranes (such as carry-deck cranes); dedicated pile drivers; service/mechanic trucks with a hoisting device; a crane on a monorail; tower cranes (such as a fixed jib, i.e., "hammerhead boom"), luffing boom and self-erecting); pedestal cranes; portal cranes; overhead and gantry cranes; straddle cranes; side-boom cranes; derricks; and variations of such equipment. However, items listed in paragraph (c) of this section are excluded from the scope of this standard.

1926.1402—Ground Conditions

(a) Definitions.

(1) *Ground conditions* means the ability of the ground to support the equipment (including slope, compaction, and firmness).

(2) *Supporting materials* means blocking, mats, cribbing, marsh buggies (in marshes/wetlands), or similar supporting materials or devices.

(b) The equipment must not be assembled or used unless ground conditions are firm, drained, and graded to a sufficient extent so that, in conjunction (if necessary) with the use of supporting materials, the equipment manufacturer's specifications for adequate support and degree of level of the equipment are met. The requirement for the ground to be drained does not apply to marshes/wetlands.

(c) The controlling entity must:

(1) Ensure that ground preparations necessary to meet the requirements in paragraph (b) of this section are provided.

(2) Inform the user of the equipment and the operator of the location of hazards beneath the equipment set-up area (such as voids, tanks, utilities) if those hazards are identified in documents (such as site drawings, as-built drawings, and soil analyses) that are in the possession of the controlling entity (whether at the site or off-site) or the hazards are otherwise known to that controlling entity.

(d) If there is no controlling entity for the project, the requirement in paragraph (c)(1) of this section must be met by the employer that has authority at the site to make or arrange for ground preparations needed to meet paragraph (b) of this section.

(e) If the A/D director or the operator determines that ground conditions do not meet the requirements in paragraph (b) of this section, that person's employer must have a discussion with the controlling entity regarding the ground preparations that are needed so that, with the use of suitable supporting materials/devices (if necessary), the requirements in paragraph (b) of this section can be met.

(f) This section does not apply to cranes designed for use on railroad tracks when used on railroad tracks that are part of the general railroad system of transportation that is regulated pursuant to the Federal Railroad Administration under 49 CFR part 213 and that comply with applicable Federal Railroad Administration requirements.

1926.1408—Power Line Safety (Up To 350 kV)—Equipment Operations

(a) Hazard assessments and precautions inside the work zone. Before beginning equipment operations, the employer must:

(1) Identify the work zone by either:

(i) Demarcating boundaries (such as with flags, or a device such as a range limit device or range control warning device) and prohibiting the operator from operating the equipment past those boundaries, or

(ii) Defining the work zone as the area 360 degrees around the equipment, up to the equipment's maximum working radius.

(2) Determine if any part of the equipment, load line or load (including rigging and lifting accessories), if operated up to the equipment's maximum working radius in the work zone, could get closer than 20 feet to a power line. If so, the employer must meet the requirements in Option (1), Option (2), or Option (3) of this section, as follows:

(i) Option (1)—Deenergize and ground. Confirm from the utility owner/operator that the power line has been deenergized and visibly grounded at the worksite.

(ii) Option (2)—20 foot clearance. Ensure that no part of the equipment, load line, or load (including rigging and lifting accessories), gets closer than 20 feet to the power line by implementing the measures specified in paragraph (b) of this section.

(iii) Option (3)—Table A clearance.

(A) Determine the line's voltage and the minimum approach distance permitted under Table A (see § 1926.1408).

(B) Determine if any part of the equipment, load line or load (including rigging and lifting accessories), while operating up to the equipment's maximum working radius in the work zone, could get closer than the minimum approach distance of the power line permitted under Table A (see § 1926.1408). If so, then the employer must follow the requirements in paragraph (b) of this section to ensure that no part of the equipment, load line, or load (including rigging and lifting accessories), gets closer to the line than the minimum approach distance.

(b) Preventing encroachment/electrocution. Where encroachment precautions are required under Option (2) or Option (3) of this section, all of the following requirements must be met:

(1) Conduct a planning meeting with the operator and the other workers who will be in the area of the equipment or load to review the location of the power line(s), and the steps that will be implemented to prevent encroachment/electrocution.

(2) If tag lines are used, they must be non-conductive.

(3) Erect and maintain an elevated warning line, barricade, or line of signs, in view of the operator, equipped with flags or similar high-visibility markings, at 20 feet from the power line (if using Option (2) of this section) or at the minimum approach distance under Table A (see § 1926.1408) (if using Option (3) of this section). If the operator is unable to see the elevated warning line, a dedicated spotter must be used as described in § 1926.1408(b)(4)(ii) in addition to implementing one of the measures described in § § 1926.1408(b)(4)(i), (iii), (iv) and (v).

(4) Implement at least one of the following measures:

(i) A proximity alarm set to give the operator sufficient warning to prevent encroachment.

(ii) A dedicated spotter who is in continuous contact with the operator. Where this measure is selected, the dedicated spotter must:

(A) Be equipped with a visual aid to assist in identifying the minimum clearance distance. Examples of a visual aid include, but are not limited to: A clearly visible line painted on the ground; a clearly visible line of stanchions; a set of clearly visible line-of-sight landmarks (such as a fence post behind the dedicated spotter and a building corner ahead of the dedicated spotter).

(B) Be positioned to effectively gauge the clearance distance.

(C) Where necessary, use equipment that enables the dedicated spotter to communicate directly with the operator.

(D) Give timely information to the operator so that the required clearance distance can be maintained.

(iii) A device that automatically warns the operator when to stop movement, such as a range control warning device. Such a device must be set to give the operator sufficient warning to prevent encroachment.

(iv) A device that automatically limits range of movement, set to prevent encroachment.

(v) An insulating link/device, as defined in § 1926.1401, installed at a point between the end of the load line (or below) and the load.

(5) The requirements of paragraph (b)(4) of this section do not apply to work covered by subpart V of this part.

(c) Voltage information. Where Option (3) of this section is used, the utility owner/operator of the power lines must provide the requested voltage information within two working days of the employer's request.

(d) Operations below power lines.

(1) No part of the equipment, load line, or load (including rigging and lifting accessories) is allowed below a power line unless the employer has confirmed that the utility owner/operator has deenergized and (at the worksite) visibly grounded the power line, except where one of the exceptions in paragraph (d)(2) of this section applies.

(2) Exceptions. Paragraph (d)(1) of this section is inapplicable where the employer demonstrates that one of the following applies:

(i) The work is covered by subpart V of this part.

(ii) For equipment with non-extensible booms: The uppermost part of the equipment, with the boom at true vertical, would be more than 20 feet below the plane of the power line or more than the Table A of this section minimum clearance distance below the plane of the power line.

(iii) For equipment with articulating or extensible booms: The uppermost part of the equipment, with the boom in the fully extended position, at true vertical, would be more than 20 feet below the plane of the power line or more than the Table A of this section minimum clearance distance below the plane of the power line.

(iv) The employer demonstrates that compliance with paragraph (d)(1) of this section is infeasible and meets the requirements of § 1926.1410.

(e) Power lines presumed energized. The employer must assume that all power lines are energized unless the utility owner/operator confirms that the power line has been and continues to be deenergized and visibly grounded at the worksite.

(f) When working near transmitter/communication towers where the equipment is close enough for an electrical charge to be induced in the equipment or materials being handled, the transmitter must be deenergized or the following precautions must be taken:

(1) The equipment must be provided with an electrical ground.

(2) If tag lines are used, they must be non-conductive.

(g) Training.

(1) The employer must train each operator and crew member assigned to work with the equipment on all of the following:

(i) The procedures to be followed in the event of electrical contact with a power line. Such training must include:

(A) Information regarding the danger of electrocution from the operator simultaneously touching the equipment and the ground.

(B) The importance to the operator's safety of remaining inside the cab except where there is an imminent danger of fire, explosion, or other emergency that necessitates leaving the cab.

(C) The safest means of evacuating from equipment that may be energized.

(D) The danger of the potentially energized zone around the equipment (step potential).

(E) The need for crew in the area to avoid approaching or touching the equipment and the load.
(F) Safe clearance distance from power lines.

(ii) Power lines are presumed to be energized unless the utility owner/operator confirms that the power line has been and continues to be deenergized and visibly grounded at the worksite.

(iii) Power lines are presumed to be uninsulated unless the utility owner/operator or a registered engineer who is a qualified person with respect to electrical power transmission and distribution confirms that a line is insulated.

(iv) The limitations of an insulating link/device, proximity alarm, and range control (and similar) device, if used.

(v) The procedures to be followed to properly ground equipment and the limitations of grounding.

(2) Employees working as dedicated spotters must be trained to enable them to effectively perform their task, including training on the applicable requirements of this section.

(3) Training under this section must be administered in accordance with § 1926.1430(g).

(h) Devices originally designed by the manufacturer for use as: A safety device (see § 1926.1415), operational aid, or a means to prevent power line contact or electrocution, when used to comply with this section, must meet the manufacturer's procedures for use and conditions of use.

Table A Minimum Clearance Distances

Voltage (nominal, kV, alternating current)	Minimum clearance distance (feet)
Up to 50	10
Over 50 to 200	15
Over 200 to 350	20
Over 350 to 500	25
Over 500 to 750	35
Over 750 to 1,000	45
Over 1,000	(as established by the utility owner/operator or registered professional engineer who is a qualified person with respect to electrical power transmission and distribution).

1926.1409—Power Line Safety (Over 350 kV)

The requirements of § 1926.1407 and § 1926.1408 apply to power lines over 350 kV except:

(a) For power lines at or below 1000 kV, wherever the distance "20 feet" is specified, the distance "50 feet" must be substituted; and

(b) For power lines over 1000 kV, the minimum clearance distance must be established by the utility owner/operator or registered professional engineer who is a qualified person with respect to electrical power transmission and distribution.

1926.1410—Power Line Safety (All Voltages)—Equipment Operations Closer Than the Table A Zone

Equipment operations in which any part of the equipment, load line, or load (including rigging and lifting accessories) is closer than the minimum approach distance under Table A of § 1926.1408 to an energized power line is prohibited, except where the employer demonstrates that all of the following requirements are met:

(a) The employer determines that it is infeasible to do the work without breaching the minimum approach distance under Table A of § 1926.1408.

(b) The employer determines that, after consultation with the utility owner/operator, it is infeasible to deenergize and ground the power line or relocate the power line.

(c) Minimum clearance distance.

(1) The power line owner/operator or registered professional engineer who is a qualified person with respect to electrical power transmission and distribution determines the minimum clearance distance that must be maintained to prevent electrical contact in light of the on-site conditions. The factors that must be considered in making this determination include, but are not limited to: Conditions affecting atmospheric conductivity; time necessary to bring the equipment, load line, and load (including rigging and lifting accessories) to a complete stop; wind conditions; degree of sway in the power line; lighting conditions, and other conditions affecting the ability to prevent electrical contact.

(2) Paragraph (c)(1) of this section does not apply to work covered by subpart V of this part; instead, for such work, the minimum clearance distances specified in § 1926.950 Table V-1 apply. Employers engaged in subpart V work are permitted to work closer than the distances in § 1926.950 Table V-1 where both the requirements of this section and § 1926.952(c)(3)(i) or (ii) are met.

(d) A planning meeting with the employer and utility owner/operator (or registered professional engineer who is a qualified person with respect to electrical power transmission and distribution) is held to determine the procedures that will be followed to prevent electrical contact and electrocution. At a minimum these procedures must include:

(1) If the power line is equipped with a device that automatically reenergizes the circuit in the event of a power line contact, before the work begins, the automatic reclosing feature of the circuit interrupting device must be made inoperative if the design of the device permits.

(2) A dedicated spotter who is in continuous contact with the operator. The dedicated spotter must:

(i) Be equipped with a visual aid to assist in identifying the minimum clearance distance. Examples of a visual aid include, but are not limited to: A line painted on the ground; a clearly visible line of stanchions; a set of clearly visible line-of-sight landmarks (such as a fence post behind the dedicated spotter and a building corner ahead of the dedicated spotter).

(ii) Be positioned to effectively gauge the clearance distance.

(iii) Where necessary, use equipment that enables the dedicated spotter to communicate directly with the operator.

(iv) Give timely information to the operator so that the required clearance distance can be maintained.

(3) An elevated warning line, or barricade (not attached to the crane), in view of the operator (either directly or through video equipment), equipped with flags or similar high-visibility markings, to prevent electrical contact. However, this provision does not apply to work covered by subpart V of this part.

(4) Insulating link/device.

(i) An insulating link/device installed at a point between the end of the load line (or below) and the load.

(ii) For work covered by subpart V of this part, the requirement in paragraph (4)(i) of this section applies only when working inside the § 1926.950 Table V-1 clearance distances.

(iii) For work covered by subpart V of this part involving operations where use of an insulating link/device is infeasible, the requirements of § 1910.269(p)(4)(iii)(B) or (C) may be substituted for the requirement in (d)(4)(i) of this section.

(5) Nonconductive rigging if the rigging may be within the Table A of § 1926.1408 distance during the operation.

(6) If the equipment is equipped with a device that automatically limits range of movement, it must be used and set to prevent any part of the equipment, load line, or load (including rigging and lifting accessories) from breaching the minimum approach distance established under paragraph (c) of this section.

(7) If a tag line is used, it must be of the nonconductive type.

(8) Barricades forming a perimeter at least 10 feet away from the equipment to prevent unauthorized personnel from entering the work area. In areas where obstacles prevent the barricade from being at least 10 feet away, the barricade must be as far from the equipment as feasible.

(9) Workers other than the operator must be prohibited from touching the load line above the insulating link/device and crane. Operators remotely operating the equipment from the ground must use either wireless controls that isolate the operator from the equipment or insulating mats that insulate the operator from the ground.

(10) Only personnel essential to the operation are permitted to be in the area of the crane and load.
(11) The equipment must be properly grounded.
(12) Insulating line hose or cover-up must be installed by the utility owner/operator except where such devices are unavailable for the line voltages involved.

(e) The procedures developed to comply with paragraph (d) of this section are documented and immediately available on-site.

(f) The equipment user and utility owner/operator (or registered professional engineer) meet with the equipment operator and the other workers who will be in the area of the equipment or load to review the procedures that will be implemented to prevent breaching the minimum approach distance established in paragraph (c) of this section and prevent electrocution.

(g) The procedures developed to comply with paragraph (d) of this section are implemented.

(h) The utility owner/operator (or registered professional engineer) and all employers of employees involved in the work must identify one person who will direct the implementation of the procedures. The person identified in accordance with this paragraph must direct the implementation of the procedures and must have the authority to stop work at any time to ensure safety.

(i) If a problem occurs implementing the procedures being used to comply with paragraph (d) of this section, or indicating that those procedures are inadequate to prevent electrocution, the employer must safely stop operations and either develop new procedures to comply with paragraph (d) of this section or have the utility owner/operator de-energize and visibly ground or relocate the power line before resuming work.

(j) Devices originally designed by the manufacturer for use as a safety device, operational aid, or a means to prevent power line contact or electrocution, when used to comply with this section, must comply with the manufacturer's procedures for use and conditions of use.

(k) The employer must train each operator and crew member assigned to work with the equipment in accordance with § 1926.1408(g).

1926.1417—Operation

(a) The employer must comply with all manufacturer procedures applicable to the operational functions of equipment, including its use with attachments.

(b) Unavailable operation procedures.
(1) Where the manufacturer procedures are unavailable, the employer must develop and ensure compliance with all procedures necessary for the safe operation of the equipment and attachments.
(2) Procedures for the operational controls must be developed by a qualified person.
(3) Procedures related to the capacity of the equipment must be developed and signed by a registered professional engineer familiar with the equipment.

(c) Accessibility of procedures.

(1) The procedures applicable to the operation of the equipment, including rated capacities (load charts), recommended operating speeds, special hazard warnings, instructions, and operator's manual, must be readily available in the cab at all times for use by the operator.

(2) Where rated capacities are available in the cab only in electronic form: In the event of a failure which makes the rated capacities inaccessible, the operator must immediately cease operations or follow safe shut-down procedures until the rated capacities (in electronic or other form) are available.

(d) The operator must not engage in any practice or activity that diverts his/her attention while actually engaged in operating the equipment, such as the use of cellular phones (other than when used for signal communications).

(e) Leaving the equipment unattended.

(1) The operator must not leave the controls while the load is suspended, except where all of the following are met:

(i) The operator remains adjacent to the equipment and is not engaged in any other duties.

(ii) The load is to be held suspended for a period of time exceeding normal lifting operations.

(iii) The competent person determines that it is safe to do so and implements measures necessary to restrain the boom hoist and telescoping, load, swing, and outrigger or stabilizer functions.

(iv) Barricades or caution lines, and notices, are erected to prevent all employees from entering the fall zone. No employees, including those listed in § § 1926.1425(b)(1) through (3), § 1926.1425(d) or § 1926.1425(e), are permitted in the fall zone.

(2) The provisions in § 1926.1417(e)(1) do not apply to working gear (such as slings, spreader bars, ladders, and welding machines) where the weight of the working gear is negligible relative to the lifting capacity of the equipment as positioned, and the working gear is suspended over an area other than an entrance or exit.

(f) Tag-out.

(1) Tagging out of service equipment/functions. Where the employer has taken the equipment out of service, a tag must be placed in the cab stating that the equipment is out of service and is not to be used. Where the employer has taken a function(s) out of service, a tag must be placed in a conspicuous position stating that the function is out of service and is not to be used.

(2) Response to "do not operate"/tag-out signs.

(i) If there is a warning (tag-out or maintenance/do not operate) sign on the equipment or starting control, the operator must not activate the switch or start the equipment until the sign has been removed by a person authorized to remove it, or until the operator has verified that:

(A) No one is servicing, working on, or otherwise in a dangerous position on the machine.

(B) The equipment has been repaired and is working properly.

(ii) If there is a warning (tag-out or maintenance/do not operate) sign on any other switch or control, the operator must not activate that switch or control until the sign has been removed by a person authorized to remove it, or until the operator has verified that the requirements in paragraphs (f)(2)(i)(A) and (B) of this section have been met.

(g) Before starting the engine, the operator must verify that all controls are in the proper starting position and that all personnel are in the clear.

(h) Storm warning. When a local storm warning has been issued, the competent person must determine whether it is necessary to implement manufacturer recommendations for securing the equipment.

(i) If equipment adjustments or repairs are necessary:

(1) The operator must, in writing, promptly inform the person designated by the employer to receive such information and, where there are successive shifts, to the next operator; and

(2) The employer must notify all affected employees, at the beginning of each shift, of the necessary adjustments or repairs and all alternative measures.

(j) Safety devices and operational aids must not be used as a substitute for the exercise of professional judgment by the operator.

(k) If the competent person determines that there is a slack rope condition requiring re-spooling of the rope, it must be verified (before starting to lift) that the rope is seated on the drum and in the sheaves as the slack is removed.

(l) The competent person must adjust the equipment and/or operations to address the effect of wind, ice, and snow on equipment stability and rated capacity.

(m) Compliance with rated capacity.

(1) The equipment must not be operated in excess of its rated capacity.

(2) The operator must not be required to operate the equipment in a manner that would violate paragraph (o)(1) of this section.

(3) Load weight. The operator must verify that the load is within the rated capacity of the equipment by at least one of the following methods:

(i) The weight of the load must be determined from a source recognized by the industry (such as the load's manufacturer), or by a calculation method recognized by the industry (such as calculating a steel beam from measured dimensions and a known per foot weight), or by other equally reliable means. In addition, when requested by the operator, this information must be provided to the operator prior to the lift; or

(ii) The operator must begin hoisting the load to determine, using a load weighing device, load moment indicator, rated capacity indicator, or rated capacity limiter, if it exceeds 75 percent of the maximum rated capacity at the longest radius that will be used during the lift operation. If it does, the operator must not proceed with the lift until he/she verifies the weight of the load in accordance with paragraph (o)(3)(i) of this section.

(n) The boom or other parts of the equipment must not contact any obstruction.

(o) The equipment must not be used to drag or pull loads sideways.

(p) On wheel-mounted equipment, no loads must be lifted over the front area, except as permitted by the manufacturer.

(q) The operator must test the brakes each time a load that is 90% or more of the maximum line pull is handled by lifting the load a few inches and applying the brakes. In duty cycle and repetitive lifts where each lift is 90% or more of the maximum line pull, this requirement applies to the first lift but not to successive lifts.

(r) Neither the load nor the boom must be lowered below the point where less than two full wraps of rope remain on their respective drums.

(s) Traveling with a load.

(1) Traveling with a load is prohibited if the practice is prohibited by the manufacturer.

(2) Where traveling with a load, the employer must ensure that:

(i) A competent person supervises the operation, determines if it is necessary to reduce rated capacity, and makes determinations regarding load position, boom location, ground support, travel route, overhead obstructions, and speed of movement neccssary to ensure safety.

(ii) The determinations of the competent person required in paragraph (u)(2)(i) of this section are implemented.

(iii) For equipment with tires, tire pressure specified by the manufacturer is maintained.

(t) Rotational speed of the equipment must be such that the load does not swing out beyond the radius at which it can be controlled.

(u) A tag or restraint line must be used if necessary to prevent rotation of the load that would be hazardous.

(v) The brakes must be adjusted in accordance with manufacturer procedures to prevent unintended movement.

(w) The operator must obey a stop (or emergency stop) signal, irrespective of who gives it.

(x) Swinging locomotive cranes. A locomotive crane must not be swung into a position where railway cars on an adjacent track could strike it, until it is determined that cars are not being moved on the adjacent track and that proper flag protection has been established.

(y) Counterweight/ballast.

(1) The following applies to equipment other than tower cranes:

(i) Equipment must not be operated without the counterweight or ballast in place as specified by the manufacturer.

(ii) The maximum counterweight or ballast specified by the manufacturer for the equipment must not be exceeded.

(2) Counterweight/ballast requirements for tower cranes are specified in § 1926.1435(b)(8).

1926.1418—Authority To Stop Operation

Whenever there is a concern as to safety, the operator must have the authority to stop and refuse to handle loads until a qualified person has determined that safety has been assured.

1926.1424—Work Area Control

(a) Swing radius hazards.
  (1) The requirements in paragraph (a)(2) of this section apply where there are accessible areas in which the equipment's rotating superstructure (whether permanently or temporarily mounted) poses a reasonably foreseeable risk of:
    (i) Striking and injuring an employee; or
    (ii) Pinching/crushing an employee against another part of the equipment or another object.
  (2) To prevent employees from entering these hazard areas, the employer must:
    (i) Train each employee assigned to work on or near the equipment ("authorized personnel") in how to recognize struck-by and pinch/crush hazard areas posed by the rotating superstructure.
    (ii) Erect and maintain control lines, warning lines, railings or similar barriers to mark the boundaries of the hazard areas. Exception: When the employer can demonstrate that it is neither feasible to erect such barriers on the ground nor on the equipment, the hazard areas must be clearly marked by a combination of warning signs (such as "Danger—Swing/Crush Zone") and high visibility markings on the equipment that identify the hazard areas. In addition, the employer must train each employee to understand what these markings signify.
  (3) Protecting employees in the hazard area.
    (i) Before an employee goes to a location in the hazard area that is out of view of the operator, the employee (or someone instructed by the employee) must ensure that the operator is informed that he/she is going to that location.
    (ii) Where the operator knows that an employee went to a location covered by paragraph (a)(1) of this section, the operator must not rotate the superstructure until the operator is informed in accordance with a pre-arranged system of communication that the employee is in a safe position.

(b) Where any part of a crane/derrick is within the working radius of another crane/derrick, the controlling entity must institute a system to coordinate operations. If there is no controlling entity, the employer (if there is only one employer operating the multiple pieces of equipment), or employers, must institute such a system.

1926.1425—Keeping Clear of the Load

(a) Where available, hoisting routes that minimize the exposure of employees to hoisted loads must be used, to the extent consistent with public safety.

(b) While the operator is not moving a suspended load, no employee must be within the fall zone, except for employees:
  (1) Engaged in hooking, unhooking or guiding a load;
  (2) Engaged in the initial attachment of the load to a component or structure; or
  (3) Operating a concrete hopper or concrete bucket.

(c) When employees are engaged in hooking, unhooking, or guiding the load, or in the initial connection of a load to a component or structure and are within the fall zone, all of the following criteria must be met:
   (1) The materials being hoisted must be rigged to prevent unintentional displacement.
   (2) Hooks with self-closing latches or their equivalent must be used. Exception: "J" hooks are permitted to be used for setting wooden trusses.
   (3) The materials must be rigged by a qualified rigger.

(d) Receiving a load. Only employees needed to receive a load are permitted to be within the fall zone when a load is being landed.

(e) During a tilt-up or tilt-down operation:

Only employees essential to the operation are permitted in the fall zone (but not directly under the load). An employee is essential to the operation if the employee is conducting one of the following operations and the employer can demonstrate it is infeasible for the employee to perform that operation from outside the fall zone: (1) Physically guide the load; (2) closely monitor and give instructions regarding the load's movement; or (3) either detach it from or initially attach it to another component or structure (such as, but not limited to, making an initial connection or installing bracing).

INDEX

A

B

C

D

E

F

G

H

I

J

L

W